Water Resources
Initiatives and Agendas

Also from Westphalia Press

westphaliapress.org

Water Resources Initiatives and Agendas

Volume 1, Number 1 of New Water Policy and Practice

Edited by Susana Neto & Jeff Camkin

WESTPHALIA PRESS
An imprint of Policy Studies Organization

Water Resources Initiatives and Agendas: Volume 1, Number 1 of New Water Policy and Practice
All Rights Reserved © 2015 by Policy Studies Organization

Westphalia Press
An imprint of Policy Studies Organization
1527 New Hampshire Ave., NW
Washington, D.C. 20036
info@ipsonet.org

ISBN-13: 978-1-63391-748-4
ISBN-10: 1633917487

Cover design by Taillefer Long at Illuminated Stories:
www.illuminatedstories.com

Daniel Gutierrez-Sandoval, Executive Director
PSO and Westphalia Press

Cheryl Walker, Development and Programs Associate
PSO and Westphalia Press

Updated material and comments on this edition
can be found at the Westphalia Press website:
www.westphaliapress.org

Table of Contents

Laudatory Greeting: Welcoming
New Water Policy and Practice Journal

Since the Royal Society in London began publishing its proceedings in the seventeenth century—and still does—the launching of academic journals by learned societies has provided a timeline of concerns and challenges. The topics reflect the times. It was only in the nineteenth century that universities began replacing their classical curriculums dominated by Greek and Latin studies with what at the time seemed new and novel disciplines. When Dr. Thomas Arnold was appointed the first history professor at Oxford, the press politely admitted being confused as to what he would find to teach, so novel seemed the idea of studying modern rather than ancient history.

We have an easier task. The subject of this journal scarcely needs justification. Whether in the suburbs of South California, or along the fracture lines of Middle Eastern boundaries, the search for enough water has become a major preoccupation of research, and a conundrum facing politicians and scientists.

This new journal then is dealing with an old problem that often enough has not been squarely faced, and which now has aggressively pushed its way into the arena of attention. The axiom that might guide us is that rather than have the answers we may need to concentrate on what are the questions. At any rate, asking the right questions and soliciting at least partial answers is the burden that falls on the editors, who have made a brave and good-humored start by soliciting board members and contributors from all over the globe.

An academic journal resembles a crossroads, with the editors charged to put up signposts and help readers find their way. One has every confidence that this journal is going to point us well and be an asset to its subject, a boost to courses and curriculum, and a worthy companion to the other Policy Studies Organization publications. Were it the launching of a ship, we would christen it with champagne. Given the topic and the prospect of its cyber journeys, perhaps one can at least make a metaphoric toast with water and wish it good speed.

Paul Rich
President, Policy Studies Organization
Garfield House, Washington DC

Editorial: Welcome to *New Water Policy and Practice Journal*

It is with great pleasure that we welcome you to *New Water Policy and Practice Journal*: a platform for the world's emerging water leaders and thinkers.

The growing global water challenges are well documented. Increasing pressure on the world´s water resources has translated into new and more interconnected challenges. In response, water management has also changed over time with the introduction of new disciplines, new techniques, new language, and new thinking. Ecology, economics, and other social sciences have all progressively added to the historic hydrology and engineering base. More recently, sociology, anthropology, and social psychology have also become part of solution to the challenges of water management. This has required re-thinking of some of the previously integrated terms and encouraged a more rigorous definition of the concepts used by different specialists. There has been a shift from seeking solutions within individual scientific disciplines to multidisciplinarity, interdisciplinary solutions and, more recently and increasingly, to transdisciplinary thinking and approaches—to seeking solutions in the interstitial spaces between scientific disciplines. There has also been a shift towards globalisation of both problems and solutions. As problems become more connected, they are also increasingly globalised. Recognition of this globalisation of water problems has triggered awareness of the opportunity to draw ideas and possible solutions from a world of practical experiences in both similar and different contexts. To address the increasingly complex water challenges the world is facing, we need transversal approaches to develop new and innovative methods that link theory and practice, management and policy.

The inspiration for *New Water Policy and Practice Journal* came from long experience and involvement in different dimensions of water management and governance, and from several highly integrated water education courses we have the pleasure of being involved in, including the Master of Integrated Water Management delivered by the International Water Centre in Australia, and the Erasmus Mundus Master of Ecohydrology delivered by a consortium of universities in Europe and South America. Recognising the excellent work coming from the participants of such courses, *New Water Policy and Practice Journal* is also dedicated to encouraging and disseminating new thinking about water policy and practice, about the knowledge and education that supports it, and about the interrelationships between them.

Sponsored by the Policy Studies Organization and supported by a dynamic team which includes a nine member International Advisory Board and over 50 editors from more than 25 different countries, *New Water Policy and Practice Journal* will be published as an online peer-reviewed open access journal, followed by hard copy, at least twice per year. Additional special issues will be published on an ad-hoc basis. Some editions will be focussed on thematic issues (e.g., urban water conflicts, water pricing policy), some on key events (e.g., international water forums and conferences, launch of new policy initiatives), some on specific exemplary management challenges and solutions (e.g., management of a particular river system), and others on emergent scientific disciplinary or education discussions (e.g., ecohydrology, new learning methodologies). All editions will feature foundational contributions from internationally recognised water leaders, together with submissions expressing new ideas from the world's young water professionals and emerging water leaders.

In this first edition of *New Water Policy and Practice Journal*, six members of our International Advisory Board give their very personal thoughts on some of the world's major water issues. With an African focus, Prof. Mike Muller from Wits University Graduate School of Governance, South Africa, and former Director General of the South African Department of Water Affairs and Forestry, begins with his thoughts on a more useful agenda for water management. Reflecting on examples of rights to water, privatisation and commoditisation of water in South Africa, the debate about dams and development in Africa, and river basins, regional institutions and state sovereignty, Mike presents the case that the current water management discourse has been distorted by strong interests with, on occasion, very significant negative impacts. Next, Dr. Zafar Adeel, Director of the United Nations University Institute for Water, Environment and Health, Canada, gives his thoughts on the role of the private sector in solving the world's water problems. Adeel opens by pointing out some of the failures of current approaches, identifies emerging opportunities that did not exist before now, describes the challenges of engaging and incentivising the private sector and finishes with the need to create positive incentives and effective regulations to support private sector contributions to resolution of the world's water problems. In the following paper, Dr. Carmen dos Santos, Scientific Coordinator Director of the Department of Biology in the University of Agostinho Neto, Angola, discusses the challenges and the recent development of the institutional framework for water management in Angola in the context of the Southern African Development Community, with particular focus on the recognition of the strategic importance of water for the regional economic integration, and the management of shared water resources. Prof. Shahbaz Khan, Deputy Director & Senior Program Specialist UNESCO Regional Science Bureau for Asia and the Pacific, Jakarta, Indonesia, then presents some ideas on how to move Integrated Water Resource Management from rhetoric to action. Reflecting on IWRM, and some of the difficulties experienced in its implementation, Shahbaz discusses some international initiatives which can strengthen IWRM in practice, identifies obstacles to operationalising IWRM principles, and proposes some opportunities to facilitate IWRM implementation at the river basin level. In the next paper, Dr. Clive Lipchin, Director of the Center for Transboundary Water Management, Arava Institute for Environmental Studies, Israel, discusses the need for decentralized approaches to transboundary wastewater management under conditions of inadequate infrastructure and political complexity, using the Israel-Palestine context in Gaza as a case example. Lastly, drawing inspiration from her work in establishing Integration and Implementation Sciences (I2S), a new discipline providing concepts and methods for conducting research on complex, real-world problems, Prof. Gabrielle Bammer from the National Centre for Epidemiology and Population Health, Australian National University, discusses the importance of documenting and communicating methods used in interdisciplinary research and in research implementation. Prof. Bammer recognises that this is a key challenge of interdisciplinary research and proposes ways that contributors to *New Water Policy and Practice Journal* can help set the direction for interdisciplinary research and communication.

With these first opinion editorials we hope that the intentions of *New Water Policy and Practice Journal* are clear and inspirational: to encourage and drive transversal approaches to the development of new and innovative methods that link theory and practice, management and policy, that address the real challenges faced by people around the world, and to provide a platform for new and emerging water leaders and thinkers to make an even greater contribution to finding the solutions so necessary to address the world's current and future water challenges.

Work relevant to *New Water Policy and Practice Journal* is happening in all countries at local, national, regional, and global scales. With an emphasis on providing a platform for new and emerging water leaders, and on sourcing ideas from around the world, we welcome contributions from research and education centres, public sector water management bodies, community groups, NGOs, private enterprises, and individuals. Submissions from young water professionals and past, present, and future students in international water education are particularly encouraged. Contributions can be in a broad range of formats, such as research papers, review papers, narratives of experiences and case studies, opinion pieces, and interviews. Written material, audio recordings, and film clips are all encouraged. *New Water Policy and Practice Journal* will provide information and announcements relevant to the journal aims, including international conferences, web-links, reports, and a calendar of events. Additionally, the Journal aims to identify particularly relevant opportunities for research funding linked to specific issue themes.

Our hope is that as water policy makers, managers, researchers, educators, community leaders, and interested individuals around the world you will find *New Water Policy and Practice Journal* a highly useful addition to the literature supporting practical solutions to water challenges.

To close, we wish to thank everyone who has offered their ideas, encouragement, and support to take *New Water Policy and Practice Journal* from a concept to this inaugural edition. To our Managing Editor, Carolina Henriques, thank you for your great efforts in helping to start the wheels turning in these challenging early stages. To all the members of our Editorial team, we thank you for the incredibly enthusiastic responses to the core concept of *New Water Policy and Practice Journal* as a platform for the world's emerging water leaders and thinkers, and for your commitment of support, which gave us the confidence to proceed with the expectation that this journal can, and will, make a substantial new contribution over time. To our International Advisory Board members we are especially grateful for the dedication to *New Water Policy and Practice Journal* that you have so clearly demonstrated already through your wholehearted and thoughtful contributions to this inaugural edition. Vitally, we wish to thank the Journal publishers, Policy Studies Organisation (PSO). To Cheryl Walker, thank you for your commitment to making things work. Finally, we are most grateful to PSO President Paul Rich and CEO Daniel Gutierrez-Sandoval for the initial invitation to develop a new water journal and for the confidence in us finding a useful niche to help address the world's growing water challenges.

Jeff Camkin   Susana Neto
Editors-in-Chief
New Water Policy and Practice Journal

A More Useful Agenda for Water Management

Mike Muller[A1]

In this opinion editorial, New Water Policy and Practice International Advisory Board member Prof Mike Muller (Wits University Graduate School of Governance and former Director General of the Department of Water Affairs and Forestry, South Africa) begins with his thoughts on a more useful agenda for water management. Reflecting on examples of rights to water, privatisation and commoditisation of water in South Africa, the debate about dams and development in Africa, and river basins, regional institutions and state sovereignty, Mike presents the case that the current water management discourse has been distorted by strong interests with, on occasion, very significant negative impacts.

Keywords: *water management; right to water; commoditization; dams and development; inter-governmental cooperation*

A review of current literature suggests that water management discourses outside of strictly technical domains are dominated by voices that are isolated from practice and, increasingly, are raised in order to promote narrow interests. Such voices have many incentives to create and maintain a divergence between theory and application, practice, and polemic.

Does this matter? To the extent that such discourse is removed from practice, it may be of little consequence. But the nature of water resources and the services derived from them means that many decision-making processes about the way that that they are developed, managed, and used—as they must be to sustain a world of nine billion people whose social and economic aspirations involve many dimensions of the resource—lie in a set of public domains that are widely accessible, at least to those with power and resources. So there is a concern—and empirical evidence in support of it—that what can be characterized as wide but weak public interests are increasingly trumped by strong, minority, and private interests.

This can pose a serious threat in many societies. Water management is a complex though not impossible task. It involves working with an often fugitive renewable resource whose availability and variability is extremely unpredictable and varies over daily, seasonal, annual, and multi-annual timeframes. This poses challenges to formal political and administrative systems that are designed to manage the predictable and measurable as well as the less formal institutions that enable communities and societies to live within the specific constraints of their water resources at local, regional, and national levels.

Although the presence or absence of water may limit the choices that can be made at any particular location, the resource itself need not be a constraint on social and economic development if broad

[A] University of Witwatersrand, South Africa

societal objectives are clear and some basic political mechanisms are in place to consider and decide on appropriate collective responses. While there is often argument about the details, it is unusual to find much dispute about the priority of providing basic water supplies for human consumption, about allowing downstream countries a reasonable proportion of a shared river's flow, or about the need to prevent pollution that will impact on communities downstream; the practical achievement of the goals is usually the larger challenge.

However, without sufficient consensus about approaches to resolve contentious issues, or effective hegemonies through which approaches can be imposed, there is a risk of dysfunction. Since the timeframes for experiencing, understanding, and intervening in response to the vagaries of water usually extend over many years, the disruption of the processes of innovation and adaptation can impose serious constraints on the functioning of a society. This is aggravated when those with the power to affect decisions have limited accountability to those whose lives they impact upon, as is often the case in what have been broadly characterized as "north-south relations" (see, for instance Doty 1996.)

In this paper, I offer a few examples from areas in which skewed scholarship, mobilized in support of activist campaigns with apparently deliberate disregard for available evidence, has demonstrably harmed large communities. And I suggest that this is becoming more rather than less prevalent. The cases presented illustrate how specific campaigns, driven by various coalitions of actors with a range of related interests, are supported by scholarship that is not just polemic but often inaccurate. It is entirely appropriate for policy and advocacy positions to be supported by research. What the examples also show, however, is

that weak scholarship has also been associated with negative outcomes for the subject communities.

Example 1: The Right To Water, Privatisation and Commoditisation

One example is the presentation of the privatization of water services as part of what is described as the commodification of water. The mere use of the terms often identifies the perspectives of the authors but that is not the issue at hand. There is undoubtedly a need to consider the nature of current economic and institutional systems for the provision of essential services and to seek more effective alternatives. And an examination of water, as a renewable, often fugitive and unpredictable, natural resource with a wide range of often competing uses can provide interesting insights into many broader issues relevant to other public services and natural resources. The limited question is whether and how such analyses affect real people in the real world. The debate in South Africa about what has been characterized as the neoliberal commodification of domestic water supplies offers a useful cautionary tale.

The issue of commodification (see Lohmann 2012 for a broader discussion of the concept) of water arose when, after much heat had been generated, it was pointed out that, contrary to the substantial literature on the subject, there was little privatization of municipal water supply and it was not really a major policy issue in South Africa (Muller 2007). Despite concerted efforts by French and British enterprises in the mid-1990s, supply was privatized in only parts of five of South Africa's 280 municipalities (of which over 150 are authorized water services authorities) and only one of those (Nelspruit) was a town of any significance. The failure of privatization to gain traction was not particularly related to opposition from trades unions and community groups as was claimed. Rather it reflected a global recognition that, contrary to initial expectations, water supply in third world cities was too complex and demanding to provide a large and profitable new market for multinational business (Budds and McGranahan 2003).

The fading of this threat to public services left a void in the campaigning arsenal of some interest groups. The threat of privatization had mobilized public service unions at national and global levels and established alliances between them and other campaign groups. Rather than declare victory and move on, there was an ongoing effort to maintain relationships between developed world campaigners and third world communities by presenting the water policy of the South African government as one that had been captured by neoliberal policies. As an apparent consequence, opposition to privatization morphed seamlessly into campaigns against "commodification". In this context, "commodifying" water was seen to entail:

- "highlighting its role mainly as an "economic good";
- attempting to reduce cross-subsidization that distorts the end-user price of water (tariff);
- insisting upon 100 percent cost recovery on operating and maintenance costs (even if capital investments are subsidized);
- promoting a severely limited form of means-tested subsidization;
- establishing shadow prices for water as an environmental good;
- solving problems associated with state control of water (inefficiencies, excessive administrative centralization, lack of competition, unaccount-

ed-for-water, weak billing, and political interference); and in the process
- fostering the conditions for water privatization". (Bond 2010)

Even this expansive and fuzzy definition was not particularly helpful for campaigners in a South African context since, rapidly following the establishment of a formal municipal financial framework, the post-1994 democratic government had regulated a "free basic water policy" which (to the displeasure of some European countries that were providing donor assistance to the sector at the time) required all municipalities to provide a basic supply of potable water free to all households and for this to be subsidized through higher tariffs for higher consumption. To sustain the attack on perceived neo-liberalism, the critics, supported by academic activists such as Bond and Dugard (2008) duly made a case that the amount of water provided was inadequate and that the means for enforcing the limit was unconstitutional. This was rejected by South Africa's Constitutional Court, which is widely regarded as a progressive trailblazer in the field of social rights (Constitutional Court 2009).

This new focus did however serve to sustain the coalition between radical academics, community-based activists (who could be predicted to support any analysis that justified an increase in the "free" allowance for their communities) and human rights campaigners. But, in making their case, the academics lost sight of broader equity issues. Specifically, since the original policy was based on the fiscal provision for municipalities and the potential for cross-subsidy, it was unhelpful to those most in need of support, the millions of rural people who were still not served by an infrastructure to provide safe water or who had to carry water from communal water points. They would not be helped by an increase in their nominal entitlement to an increased volumetric amount of water to levels greater than could practically be drawn. But they would also likely suffer from the reallocation of financial resources to meet the louder and more visible urban demands.

Even in the urban areas, poor people have not benefited greatly from the campaign to increase the volume of free water provided. Municipal leaderships, recognizing that their budgets could be threatened by the need to provide an ever-expanding amount of water, sought options. Conservative policy advisors, who were offended by free water going to the "non-poor" and opposed the general free allocation, were obvious allies. The predictable outcome of this coalition was that many municipalities have introduced means-testing and offered free basic water only to "indigents." Means tests are well known to exclude theoretically eligible recipients of benefits, because of the transaction challenges that they pose, which include exposing poor and vulnerable people to new modes of administrative corruption. Also lost was the broader principle that had been established, which was that all members of a community should have access to a basic amount of water, a provision that specifically sought to "decommodify" basic water supply. And the Constitutional Court judges explicitly commented that ".... the applicants proposed no third way as an alternative to the provision of universal benefits or means-tested benefits" (Constitutional Court 2009).

Aside from the impact on the poor, these campaigns have had an impact on scholarship and policy. Other authors uncritically take as fact assertions that are repeated sufficiently often even though they are not referenced against any formal source but merely repeated; as a result, many re-

searchers continue to report that South Africa's water supplies were being privatized on a large scale, that wildly exaggerated numbers of people were being cut off for non-payment, and that the amount of water provided was not adequate (although it complied with World Health Organization guidance). The misinformation provided also distracted policy-makers from the fact that the main cause of lack of access to safe water in a country which has provided nominal access for 95% of its population are management failures, aggravated by extreme decentralization and autonomy for local authorities, which was a product of South Africa's negotiated settlement in 1994.

The activist campaigns and the scholarship that underpins them have thus been unhelpful across a variety of dimensions. As I wrote at the time,

"For the minority, the practitioners who simply seek to achieve the broad goals of service delivery, it remains important to develop an understanding of the larger politics so as to be able to promote the interests of the communities they seek to serve as effectively as possible. There continues to be a need, if not for parish pump politics, at least for a politics that will help communities to keep their parish pumps working. That is a requirement of a modern society, without which many of the higher goals are unlikely to be achieved." (Muller 2007)

Example 2: Dams and Development

The dams and development debate was promoted by a coalition of strong voices from northern environmental groups working with southern social activists to oppose the construction of large dams. There have undoubtedly been many cases in which large dam construction projects have been associated with the displacement and impoverishment of local people in developing countries. However, these impacts are not specific to dams. They occur perhaps more frequently and extensively with road, rail, mining, and commercial agriculture projects as well as with private property development and urban expansion more generally.

However, the literature generated by the dams discourse has generally avoided consideration of the more generic issues of expropriation, or eminent domain as it is known in the United States, or of the political economy of large public and private projects in different contexts. This reflects the primary objective of the campaigns which was to prevent the construction of large dams rather than to promote equitable development and social justice. However, the joint focus created a coalition which provided social legitimacy for the environmentalists and put the formidable lobbying capacity of the environmentalists at the disposal of the social activists. In this context, the distinction between academic research and polemic is fuzzy, to say the least, and the "scholarship" underpinning these campaigns is often thin.

Analysis of this case is helped by the fact that two of the main actors have provided detailed personal accounts of their objectives and methods and reflected on the outcomes (McCully 2003; Briscoe 2010a; 2010b). In his unusually frank disclosure, Patrick McCully, then campaigns director of the International Rivers Network provided insight into the campaigners' objectives, writing

"... as someone involved in the process from the beginning. I coordinated efforts by dam critics to lobby for an independent international review committee of dam building and partic-

ipated in decisions regarding the committee's composition and mandate. I have also served as coordinator of the ICDRP since its inception in May of 1998. Thus, this paper reflects a personal perspective on how the participation of activists resulted in a Commission with the ability to deliver favorable results."

Among the favorable results he referred to was the way in which the World Commission on Dams was established as

"... a globalized and privatized policy process. The public sector was, to a significant extent, marginalized from the process, and much of its accustomed political space taken up by civil society and the private sector."

... anti-dam activists saw the WCD as a means to further the aims of the international movement against dams by getting a thorough and unbiased review of the actual impacts and performance of dams that would be difficult for dam promoters to ignore or discredit. Dam critics realized that it was extremely unlikely that a multi-stakeholder commission would take a firm "no dams" stance. But they correctly believed that such a commission could set strict criteria for future dams, that, if followed, would prevent most destructive dams from going forward, promote better alternatives, and help promote recognition of the need for reparations for past damage due to dam construction. To adapt Clausewitz's famous dictum, the WCD was a mere of the anti-dam struggle by other means.

What are the outcomes? It is difficult to generalize, beyond the demonstrable conclusion of Briscoe that lending for large water resource infrastructure projects by Multilateral Development Finance Institutions such as the World Bank declined substantially. But impacts can also be demonstrated at a more local scale with individual projects. A useful example, because it has been well publicized and is relatively well documented, is that of the Bujagali dam in Uganda whose construction was delayed for over five years by northern-based environmental activists in which McCully's IRN played a leading role.

The dam was the intervention chosen at the turn of the century as the best alternative for Uganda to meet its growing power needs. It has a small physical footprint because it is essentially a run-of-river installation, a few kilometers downstream of the existing hydropower installations at Owen Falls, which regulate the flow from Lake Victoria into the Nile River. Claims by campaigners that thousands of people would be displaced by the project turned out to include mainly those displaced by power line construction rather than the dam itself. Just a handful of local farmers used land in the few hundred hectares of the valley to be flooded, although some households were relocated to provide land to establish construction and administration facilities.

When the project was initiated in 2001 with a target date for completion of 2005, Uganda was already facing critical energy shortages. The project was delayed by a succession of procedural complaints and only finally commissioned in 2012. Already, in 2007, the World Bank had noted that

"…… if the Bujagali project had been successfully financed in 2002, Uganda would have been able to avoid the current economic penalties. Moreover, the reductions in Lake Victoria water levels from over-abstraction for hydropower production may not have occurred. This is because the Bujagali project is downstream of the current Nalubaale/ Kiira dam complex, and will re-use the upstream water releases. When commissioned, the proposed project will produce power at a fraction of the cost Uganda is now paying for the supply from thermal power plants running on imported fuel." (World Bank 2007)

By the time the project was finally commissioned, seven years late in 2012, the considerable damage done to the Ugandan economy had been well documented. With the insights provided by Briscoe 2010b, the delay could clearly be attributed to externally driven anti-dam campaigns. In turn, the delay in meeting the country's electricity needs contributed to slower economic and employment growth (See, for instance, IMF 2007; Uganda BoS 2011). Given the demonstrable links between unemployment and poverty (wage earners' incomes were significantly higher than subsistence incomes (Ellis and Bahiigwa 2003)) and between poverty and child mortality ("Infant mortality is found to be almost 80% higher for the poorest 20% compared with the richest 20%" (MFPED 2002)), it can be concluded that the delays in completing Bujagali contributed to some thousands of additional child deaths in the country.

Can the achievements of the campaigners be balanced against childrens' lives? Were the lives lost perhaps outweighed by benefits, such as a more careful approach to dam building, accrued in other cases? This calculus is unlikely ever to be made; the point however is that it cannot be assumed that campaigns of this nature and the scholarship that is mobilized to support them are without consequence.

The opposition to dams has developed a number of lines of critique aside from specific allegations of corruption, displacement of people without compensation, and environmental damage. A particular focus has been to exclude dams for hydropower generation from sources of funding for renewable energy. This has been based on claims that hydropower dams emit methane at a rate that exceeds (in terms of warming impacts) the benefits in terms of emission reductions from fossil fuels foregone (see for instance Giles 2006). This hypothesis has been used to support campaigns which have significantly constrained finance for hydropower projects in developing countries that are eligible for subsidies under the Clean Development Mechanism (CDM). Yet, despite pressure from environmental organizations, the Intergovernmental Panel on Climate Change (IPCC), the global forum established to provide science-based advice to guide climate policy formulation, had long concluded that

> "While some GHG emissions from new hydroelectric schemes are expected in the future, especially in tropical settings, in the absence of more comprehensive field data, such schemes are regarded as a lower source of CH4 emissions compared to those of other energy sector or agricultural activities. Hydroelectric power is therefore not treated as a separate emission category..." (IPCC 2000)

The exclusion of hydropower from CDM funding has continued even though the IPCC later reported that more detailed investigation simply confirmed that "for most hydro projects, life-cycle assessments have shown low overall net GHG emissions"(IPCC 2007), a position that has been maintained in the IPCC's 5th Assessment Report in 2014.

In this case, anti-dam lobbies have constrained the development of power generation opportunities that would have reduced the emissions of global warming gases. The effect of this has been limited in countries such as China and Brazil, which do not depend on external financing and its conditionalities. The impact has been greatest on poorer countries in Africa and South East Asia which have been denied access to cheaper, "cleaner" energy by externally imposed preferences.

Example 3: River Basins, Regional Institutions and State Sovereignty

The final example of the way in which water management agendas have been promoted through the mobilization of directed scholarship is that of the approach to the governance and management of water of "transboundary" rivers that flow through more than one country. In this case, northern governments have aggressively promoted an agenda of cooperation on transboundary water management in developing countries, with the encouragement of northern environmental activists.

While this could be seen as a laudable contribution to development, I have argued elsewhere (Muller 2012; Muller 2011) that the approaches taken are inimical to African interests and reflect primarily northern environmental protection goals rather than the achievement of effective cooperation between countries to facilitate better water management in support of sustainable development. This has occurred, in part, because Africa's relatively weak regional institutions have been used as entry points to promote policies and activities that would not be adopted if individual governments had fund them from their own budgetary resources. Aside from the sub-optimal use of Africa's scarcest resource—trained technical human resources—negative outcomes have included the failure of Southern Africa's efforts at practical regional cooperation in the important field of energy and serious electricity shortages.

It is a matter of fact that many of Africa's rivers are shared between more than one country in a variety of geographical configurations. This is cited as the potential cause for conflict. Further, it is often stated (and taken for granted) that the development of water resource infrastructure on shared rivers should be undertaken jointly by riparian states. To this end, it is recommended that River Basin Organizations be established by riparian countries and be given sovereign powers to undertake such development and manage the water.

These recommendations appear reasonable and more or less consistent with practice in other infrastructure sectors where cooperation is necessary to join roads, railways, and energy grids. Yet it ignores the fact that the shared rivers already provide the infrastructure linkages that other sectors need to build. And the assumption that joint projects are inherently preferable is often incorrect because Africa's water resources are (with notable exceptions in the south and north of the continent) still barely used. There are thus usually better national projects (in terms of cost and location) than projects shared between a number of countries—and that is before the transaction costs of cooperative projects are considered (this is documented in more detail in Muller 2014). What is required on shared rivers is some system of communication to ensure that information about hydrology and other dimensions can be shared and proposed developments and their cross-border impacts can be discussed.

It has been repeatedly demonstrated in practice that the institutional requirements for this can usually be met by inter-governmental committees rather than stand-alone institutions. Indeed, evidence from other parts of the world including Europe (with the Rhine) and South East Asia

(the Mekong) suggest that areas of water management that require the involvement of national governments are best organized on a cooperative rather than an integrated basis, in order to ensure effective liaison between sectors within each country which is often more challenging than the liaison between water authorities of different countries. In Africa, almost all effective river basin cooperation that has produced demonstrable outcomes has been organized on the basis of ad hoc inter-governmental cooperation rather than through formal River Basin Organizations.

Yet the River Basin Organization is presented as best practice in much of the literature. As Mukhtarov and Gerlak (2013) point out, it is not clear why this should be so:

"…. the role of individual and collective actors in advancing and maintaining RBOs is relatively unknown (Molle 2009, 484). We understand little of the 'how and why' discourse and policy initiatives have emerged in this context (Mollinga et al. 2006: 30), and what role transnational actors play in the promotion of such policies as 'best practices.'"

It is suggested that much of the policy priority given to this form of organization lies in the fact that, for environmental conservation and management, the river basin is a natural unit of water management, the river constituting the aquatic ecosystem in its own right. It is true that river basins are important units of analysis in assessing physical water availability and conditions. However water users often work with very different spatial perspectives which are both larger and smaller than the river basin. They may operate within the jurisdiction of a local or state government which sets effluent quality standards or simply as a member of a group of farmers who share access to a canal. Large urban areas often obtain their water supplies from beyond the basin in which they lie and power for many purposes may be generated from dams on rivers that lie in other countries. So the adoption of the river basin as a geographical scale of management is not an obvious approach, although not all commentators would be as blunt as Graefe (2011) who concludes that

"The river basin fetishism, the domination of the IWRM and governance concepts can be taken as a symptom of the depolitization of water management. It has to be understood as an effort to create new environmental regions voided of political interests, political representations and overall of politics."

Some of the mechanisms by which particular approaches are promoted in poor countries are self-evident. The ability of donor countries to use the hegemony afforded them as major financial supporters of many African governments to promote policy positions has been systematically exploited by environmental activists. A particularly egregious recent example of this is afforded by the introduction in the U.S.'s budget law of a provision in relation to the financing of dams by international development banks. Senator Leahy from Vermont, a state whose residents depend on hydropower but have little connection with the third world, introduced an amendment into the 2014 U.S. budget, now passed into law, which stipulated that:

"The Secretary of the Treasury shall instruct the United States executive director of each international financial institution that it is the policy of the United States to oppose any loan,

grant, strategy or policy of such institution to support the construction of any large hydroelectric dam." (Section 7060(c)(7)(D).) (USA 2014)

Senator Leahy is well known as a supporter of environmental organizations which, in turn, provide him with an important supportive constituency. There is no evidence that he has considered the merits of the case or how he proposes to account to those citizens of other countries whose decisions he is so gratuitously usurping. This example highlights the role of international environmental non-governmental organizations (NGOs) and the extent to which they act in alliance with rich world governments.

So there is indeed reason for concern when academic writers suggest that

"….. fresh water, its availability and use should now be recognized as 'a common concern of humankind', much as climate change was recognized as a 'common concern of humankind' in the 1992 United Nations (UN) Framework Convention on Climate Change (UNFCCC 1992) and conservation of biodiversity was recognized as a 'common concern of humankind' in the 1992 Convention on Biological Diversity (CBD)." (Weiss 2012)

But there are also more subtle forms of hegemony, not least the funding and dissemination of research. An example is the evolution and theorization of transboundary environmental governance in the water space as a focus for global environmental organizations. Unpacked, this is an attempt to specify transboundary rivers as sites that require common management structures and to remove their oversight from the purview of national governments. Academic engagement is one channel through which this approach is promoted.

In this context, the drive by environmental non-governmental organizations to promote River Basin Organizations is understandable, just as it is unhelpful for the many other dimensions of water management. This also explains the reticence of the majority of developing countries when deciding whether to ratify instruments such as the UN Convention on Shared Rivers. This was approved in 1997, but despite pressure from international environmental organizations such as WWF, was only ratified by the 35 countries needed for it to come into force in June 2014, 17 years later (UN 2014). Many of the countries that have ratified have no significant international rivers; the 35th ratification, by Vietnam, should be seen as riposte to China for its oil exploration in contested parts of the South China Sea rather than as a water management decision, since other countries sharing the Mekong river have not yet ratified.

This helps us to answer the question, "does this matter"? In Southern Africa at least, there is already evidence that the strategies used to assert regional environmental sovereignty in water management and constrain national discretion have done damage, by commission and omission. Apparently specific instruments such as the RAMSAR Convention (the "Convention on Wetlands of International Importance, especially as Waterfowl Habitat") can and have been used to prevent countries from using water from rivers that flow through their territory; this happened to Namibia with the Okavango River (Ramberg 1997), constraining the country's ability to respond to the impacts of a severe drought.

An example of the impact of omission, also from Southern Africa, is the failure of the region to respond to an emerging electricity supply crisis. Over the first decade of the twenty-first century, the growth in economic activity saw a parallel growth in electricity consumption. Although this was noted by the electricity authorities at a national level, and discussed by electricity specialists at a regional level, national and regional water agencies could not attract the funding required to promote the preparation of hydropower projects in time to meet the demand. In consequence, the demand is being filled by two massive, emission-intensive three 600 MW coal-fired power stations, the opportunity for local development and regional cooperation has been lost and the region has suffered five years of crippling power shortages.

Discussion: Consciously Building the World's Anthropocene Future

This review has sought to provide some examples of how the current water management discourse has been distorted by strong interests with, on occasion, negative impacts on significant communities. The purpose of establishing this is to make the case for new approaches, which can better serve the larger, if weaker set of what could loosely be described as public interests. (The concept here of strong and weak interests refers to the resources which relevant interest groups are willing and able to mobilize to achieve specific sets of goals and the priority that they give to these goals.)

The purpose is not to propose a new paradigm for water management. Indeed, the issues presented here show that it is unlikely that a single theoretical paradigm will be able to address the multitude of diverse water-related issues that confront different communities. It is suggested that what is required is a more balanced approach with greater attention to and respect for the empirical and less for the ideological and polemical.

In part, the distortions observed in the discourse are simply the consequence of a difference of context and perspective. Commentators from the essentially static societies of Western Europe and North America, still dominate the literature in output if not in quality or relevance. In these countries, most of the physical and institutional development required to meet water security needs has already occurred. There are decisions to be made as infrastructures are upgraded and replaced and institutions adapted to fit the political and economic logic of the moment. But these decisions are at a different order of magnitude to those confronted by the practitioners of the South who face the very different challenges of meeting the needs and wants of societies where urban populations have doubled in size in the past 25 years and will probably double again by 2050; where social expectations double the requirements for water yet again; where still more is required to support economic activity along with more investment to manage the growing deluge of used water that is produced; where the financial resources to support fully this rapid growth are rarely available; and where the political economy is more complex, institutions more fragile, and capabilities more scarce.

The fact that so many authors write from an easier and very different context to that in which the majority of the world's population lives means that the assumptions and preferences (explicit and implicit) that guide the analysis are also different. This helps to explain the general priority given to environmental protection and conservation that colours much analysis. Developing countries face different challenges and have different priorities. They have not yet established the stable physical geography of a predominantly urbanized community and adjusted their landscapes to support it, as has happened in the richer world. Their priority (again, explicit or implicit) must be to build a geography that sustainably supports their popula-

tions. Given population numbers, the scale of such interventions will inevitably impact substantially on the "natural" environment. But that is not helpful terminology. The present is best considered as the Anthropocene moment. People now determine the characteristics of their future "natural" environment, from its atmosphere and its waterways to its ground covers and the biota that inhabit them. They need the science and analysis to support that adventure not research that wistfully seeks to protect the past. This need is not yet generally reflected in the water discourse.

Tony Allan, originator of the virtual water concept, has highlighted the challenge:-

> Policy debates bring about hegemonic convergence, a concept, which is similar to that of sanctioned discourse. Both terms are part of a political ecology approach to water policy making and help to show how environmental policy-making is made. (Hajer 1996) All policy making discourse is partial in that it is made by coalitions, which reflect those who can best construct and deliver the most persuasive arguments. The most persuasive can exclude the voices of those who do not construct their messages sufficiently well to gain access to the discourse. Policy outcomes are the result of elites making deals selectively with groups that cannot be gainsaid. (Allan 2003)

In the context of the dams and development debate, another long-time analyst of the water sector noted that the divergence between the proposals from the rich world and the needs of the poor:-

> "... contributed to a concerted action by the developing countries which

were forced to unite by the biased report which otherwise may not have happened. With a combined voice, they could tell developed countries who had already constructed most of their large dams, that infrastructure construction is important for their socio-economic development and that they need such structures to produce food, generate energy employment and income, provide basic services and improve the overall quality of life of their citizens." (Biswas 2012)

In the same context, Briscoe referred to the WCD process which

".... ended up as an audacious attempt by NGOs to impose policies on governments and inter-governmental institutions. The overreach was so great that even normally placid governments reacted and the core, the WCD guidelines, were rejected by all governments building dams and by all IFIs. (A few rich country governments where green parties are strong, notably Germany, remain doggedly committed to the WCD.)."

A useful response would see the emergence of a different focus for research. Mollinga and Gondhalekar recently noted the continued impact of the long-standing separation of natural and engineering sciences from social sciences and the humanities. They highlight the impact of the now-growing interaction through its association with the modernization of international development efforts, which has indeed provided the paradigm and set the tone for much research on aspects of water. They suggest a new and more rigorous comparative approach to water research. (Mollinga and Gondhalekar 2014)

Such approaches will require some introspection on the part of the research community, in particular, the extent to which it provides what could (dangerously in the context of modern social science) be called useful and objective analysis. It might help if there was more interplay between researchers and practitioners. Briscoe laments the disconnect between them:-

"I see this disconnect between those who opine and advise (frequently with no practical knowledge, and usually for others to live with the consequences) and those who do and know to be a dangerous gap in this water-aware world. The truth is that informing an ever-more-interested public is a vital task. And the sad truth is that those who opine are much more effective at dealing with the media (and frequently those who define the agenda for international financing institutions) than those who do. Therein lies, I suppose, the germ of a discussion for another day, perhaps one that can help bring together a 'coalition of those who do'!" (Briscoe 2010a).

That sentiment is increasingly often expressed. Even some of the authors of the more ideological and polemical literature cited here have recognized the need for an approach that is perhaps more empirically focused. In support of the arguments raised against commoditization some of these writers have suggested that forms of management, under "community" control, would offer better alternatives. In this context, the work of Elinor Ostrom is often cited as evidence that the "tragedy of the commons" is not an inevitable outcome. But a researcher like Bakker, who has generated a substantial literature on the subject has also expressed concerns:

2007).

"…. caution is also merited, insofar as appeals to the commons run the risk of romanticizing community control. Much activism in favour of collective, community-based forms of water supply management tends to romanticize communities as coherent, relatively equitable social structures, despite the fact that inequitable power relations and resource allocation exist within communities . Although research has demonstrated how cooperative management institutions for water common pool resources can function effectively to avoid depletion other research points to the limitations of some of these collective action approaches in water." (Bakker 2007)

From Mcdonald and Ruiters comes the reflection that

In thinking about alternatives, it became clear that a systematic study of concrete practices needed to be grounded in principles—objective realities of uneven development and varying political circumstances in diverse parts of the world—not just a random collection of case studies. Visions and principles require more than a mere summary of what activists and progressive policy makers have to say. They require a historical and multiscalar view of links between public services and democratic governance. And while activ¬ists can share stories of oppression and repression, and distil principles of justice and visions for alternatives from their own experiences, this needs to be complemented by analysis and synthesis informed by understandings of market crises. (McDonald and Ruiters

The U.S. West Coast can be regarded as a sentinel site for conflicts between environmental and development values. In that context, the comments by fisheries professor and EPA scientist Robert Lackey are relevant. He argues that

"…. unless we are more vigilant guarding against the misuse of science in natural resource policy and management … we risk marginalizing the helpful role that science and scientists can play in resolving important, but divisive natural resource issues." (Lackey 2009)

He illustrates this by the persistent misuse of terms such as "environmental integrity" which serve to colour debates that are really about societal values and preferences. He notes that, in discussions about biological diversity, alien species are routinely excluded; the apparently preferred goals of environmental restoration often conceal the real choice which is between two forms of altered environment. As he explains:

"In a democracy, having widely available, accurate, understandable, and unbiased scientific information is central to the successful resolution of the typically contentious, divisive and litigious natural resource policy issue. To allow science to be marginalized through misuse is a major loss to society and its decision making institutions."

Of course, many post-modernists would query much of that statement, and in particular the nature of the decision-making institutions and the discourses that they use. But in doing so, they too undermine the potential of such institutions, flawed though it may be, to serve a broader set of interests.

They fail to offer practical alternatives and, in so doing, strengthen the hands of those who would make choices by distorting discourses to enforce their own preferences rather than recognizing the claims of a wider society.

[1]*Disclosure of Interests from the Author*

Since I am critical of other authors and their interests in this paper, I should declare my own position. I come to this topic from a variety of perspectives, as a practitioner active at both operational and policy levels in water resources and water services, from local to regional scale in southern Africa, particularly Mozambique and South Africa; as a Commissioner with South Africa's National Planning Commission and therefore still an actor in some of the issues; I have been an activist on health and development issues, including campaigns against harmful marketing in developing countries by milk, pharmaceutical and tobacco multinationals; I also wrote on environment and development issues for Earthscan in the 1970s. But as a student and a researcher (inter alia, as visiting adjunct professor at the University of Witwatersrand School of Governance), I also try to understand some of the technical dimensions of a range of critical policy issues in order to inform those debates. As a practitioner and activist, I am conscious of the importance of objective scholarship and unbiased information which can support robust societal processes to make the best possible decisions about contentious issues. After extensive engagement with international aid community including six years with the Global Water Partnership's technical advisory committee and two years as chair of a World Economic Forum "Agenda Council" on water security, I have argued that much of the discourse about water that is imposed on developing countries, initially through aid-driven hegemonies but increasingly through other channels, is actually damaging their people and economies rather than helping them to meet their diverse societal goals.

References

Allan, T. 2003. "IWRM/IWRAM: A New Sanctioned Discourse?" Occasional Paper 50 SOAS Water Issues Study Group School of Oriental and African Studies/King's College London University of London, April 2003.

Bakker, K. 2007. "The 'Commons' versus the 'Commodity': Alter-Globalization, Anti-Privatization, and the Human Right to Water in the Global South." *Antipode* 39 (3): 430-455.

Biswas, A. 2012. "Impact of Large Dams: Issues, Opportunities and Constraints." In *Impacts of Large Dams: A Global Assessment*, eds. Cecilia Tortajada, Dogan Altinbilek, Asit Biswas. Springer. 2012.

Bond, P. 2010. "Water, Health, and the Commodification Debate." *Review of Radical Political Economics* 42 (4): 452.

Bond, P. and Dugard, J. 2008. "The Case of Johannesburg Water: What Really Happened at the Pre-Paid 'Parish Pump.'" *Law, Democracy & Development*, 12 (1): 1-28.

Briscoe, J. 2010a. "International Financing Institutions and Hydropower in the Developing World." *Hydropower & Dams* 17 (6): 55-59.

Briscoe, J. 2010b. "Viewpoint—Overreach and Response, The Politics of the WCD and Its Aftermath." *Water Alternatives* 3 (2): 399-415.

Budds, J. and McGranahan, G. 2003. "Are the Debates on Water Privatization Missing the Point?" *Experiences from Africa, Asia and Latin America, Environment and Urbanization* 15 (2): 87-114.

Constitutional Court. 2009. Judgement. Mazibuko and others vs City of Johannesburg and others, Case CCT 39/09, 2009 ZACC 28.

Doty, R.L. 1996. *Imperial Encounters: The Politics of Representation in North-South Relations*. Minneapolis, University of Minnesota Press.

Ellis, F. and Bahiigwa, G. 2003. "Livelihoods and Rural Poverty Reduction in Uganda." *World Development* 31 (6): 997–1013.

Giles, J. 2006. "Methane Quashes Green Credentials of Hydropower." Nature 444, No.3(2006): 524-25.

Graefe, O. 2011. "River Basins As New Environmental Regions? The Depolitization of Water Management." *Procedia Social and Behavioral Sciences* 14: 24-27.

Hajer, M. 1996. The Politics of Environmental Discourse: Ecological Modernization and the Policy Process. Oxford: Clarendon Press.

IMF. 2007. Uganda: 2006 Article IV Consultation and Staff Report for the 2006 Article IV Consultation, Country Report No. 07/29.

IPCC. 2000. "Intergovernmental Panel on Climate Change." Special Report on Emissions Scenarios. WMO/UNEP.

IPCC. 2007. IPCC Fourth Assessment Report: Climate Change (ch4 s4.3.3.1 hydroelectricity), WMO/UNEP.

Lackey, R.T. 2009. "Is Science Based Towards Natural?" *Northwest Science* 83 (3): 277-279.

Lohmann, L. 2012. "Performative Equations and Neoliberal Commodification: The Case of Climate." In *Nature Inc.: Environmental Conservation in the Neoliberal Age*, eds. Bram Buscher, Wolfram Dressler, Robert Fletcher. University of Arizona Press, 29 May 2014 http://www.thecornerhouse.org.uk/sites/thecornerhouse.org.uk/files/Performative%20Equations5.pdf.

McCully, P. 2001. "The Use of a Trilateral Network: An Activist's Perspective on the Formation of the World Commission on Dams." American University International Law Review 16, no. 6 (2001): 1453-1475.

McDonald, D. and Ruiters, G. 2007. "Ways Forward for Alternatives in Health Care, Water and Electricity." *In The Age of Commodity: Water Privatization in Southern Africa*, eds McDonald & Ruiters.

MFPED (Ministry of Finance, Planning and Economic Development). 2002. Infant Mortality in Uganda 1995-2000 Why the non-improvement? Discussion Paper, 6 August 2002

Molle, F. 2009. "River-Basin Planning and Management: The Social Life of a Concept." *Geoforum* 40: 484-494.

Mollinga, P. and Gondhalekar, D. 2014. "Finding Structure in Diversity: A Stepwise Small-N/Medium-N Qualitative Comparative Analysis Approach for Water Resources Management Research." *Water Alternatives* 7 (1): 178-198.

Mollinga, P., Dixit, A and Athukorala, K. (eds). 2006. Integrated Water Resource Management: Global Theory, Emerging Practice and Local Needs. SAGE Publications, New Delhi, December 2012.

Mukhtarov, F. and Gerlak, A. 2013. "River Basin Organizations in the Global Water Discourse: An Exploration of Agency and Strategy." *Global Governance: A Review of Multilateralism and International Organizations* 19 (2): 307-326.

Muller, M. 2007. "Parish Pump Politics: The Politics of Water Supply in South Africa." *Progress in Development Studies* 7 (1): 33–45.

Muller, M. 2011. "The Political and Practical Challenges of Designing and Implementing an African Water Resource Management Agenda." *In Africa in Focus: Governance in the 21st Century,* eds. Kwandiwe Kondlo, and Chinenyengozi Ejiogu. HSRC Press, February 2011, Capetown.

Muller, M. 2012. Asymmetry and accountability deficits in water governance as inhibitors of effective water resource management, (in) Global Water Systems Project: *River Basins and Change*, eds. Janos J. Bogardi, Jan Leentvaar and Hans-Peter Nachtnebel. GWSP IPO, Bonn, 194-198.

Muller, M. 2014. *A Program to Support Transboundary Water Management and Regional Integration in Africa*. Tunis: African Development Bank.

Ramberg, L. 1997. A pipeline from the Okavango River? Ambio 26(2):129.

Uganda BoS. 2011. Statistical Abstract 2011, Uganda Bureau of Statistics.

UNFCCC. 1992. United Nations (UN) Framework Convention on Climate Change. New York, NY (US), 9 May 1992, available at: http://unfccc.int.

UN (2014), United Nations Treaty Collection, Status as at 22-09-2014, Chapter XXVII, Environment, 12. Convention on the Law of the Non-Navigational Uses of International Watercourses (accessed at) https://treaties.un.org/Pages/ViewDetails.aspx?src=IND&mtdsg_no=XXVII-12&chapter=27&lang=en

USA 2014. Consolidated Appropriations Act of 2014 (Section 7060(c)(7)(D).), Government of the United States, Washington.

Weiss, E.B. 2012. "The Coming Water Crisis: A Common Concern of Humankind." *Transnational Environmental Law* 1: 153-168.

World Bank. 2007. *Uganda—Private Power Generation Project (Bujagali)*. Washington, DC: World Bank. http://documents.worldbank.org/curated/en/2007/03/7491912/uganda-private-power-generation-project-bujagali.

Engaging and Incentivizing the Private Sector: An Emerging Opportunity for the Water World

Dr. Zafar Adeel[A]

In this opinion editorial, New Water Policy and Practice International Advisory Board member Dr Zafar Adeel (Director of the United Nations University Institute for Water, Environment and Health, Canada) gives his thoughts on the role of the private sector in solving the world's water problems. Adeel opens by pointing out some of the failures of current approaches, identifies emerging opportunities that did not exist before now, describes the challenges of engaging and incentivising the private sector and finishes with the need to create positive incentives and effective regulations to support private sector contributions to resolution of the world's water problems.

Keywords*: private sector; water governance; privatization; regulation; incentives*

Constraints of Business-as-Usual Approaches

We all know the well-recited figures about access to water and sanitation: the latest JMP report cited about 700 million people without access to an improved source of water (WHO and UNICEF, 2014). However, the actual situation on the ground is even worse: the number of people without safe drinking water is about two billion (WWAP, 2014). Concealed beneath this lack-of-access-to-a-critical-service story lies a major societal driver that impinges on human health, social development, economic productivity, primary education, and livelihood opportunities. Failure to address this challenge, therefore, can effectively reverse development gains made in many developing countries and block the achievement of new goals.

The effective management of water resources, particularly as water availability shrinks in many water-scarce regions of the world due to climatic changes and as water quality diminishes due to enhanced urban and industrial activities, also remains a major challenge (WWAP 2012). While a UN-Water report highlighted that over 80% of the countries have undertaken reforms to improve water management (UNEP 2012), the situation in most developing countries remains far from satisfactory. There are three key gaps that limit most developing countries' ability to respond to their water resource management challenges: shortage of adequately trained and qualified human resources, lack of access to appropriate and adequate technological capacity, and insufficient allocation of financial resources.

There is also the specter of transboundary water-sharing conflicts that refuses to go away. While the myth of 'water-wars' has been debunked through research and recent history, most recent studies point to the challenges of effectively managing wa-

[A] United Nations University, Canada

ter resources that cut across national and sub-national boundaries (UN-Water 2013). An interesting report from the U.S. intelligence community talks about the potential failure of states that are ill equipped to handle their water woes within their borders and with their neighbors (NIC 2012). The report also points to food security and management of irrigation waters as being closely linked to the broader notion of water security.

In the backdrop of these challenges in providing water-related services, managing scarce water resources and maintaining water security, the international and national responses have yielded limited success. The JMP reports that nearly two billion people gained access to an improved source of water between 1990 and 2010 (WHO and UNICEF 2014). And yet the number of deaths tied to water-related diseases does not show any appreciable decline or a decreasing trend in this period; perhaps we need to consider how safe the "improved" water sources really are for the health of those consuming that water. Many countries in Sub-Saharan Africa show regression in provisioning of water and sanitation services as population pressures get larger. In contrast, the water storage and hydropower generation potential for this region also remains largely untapped (WWAP 2014).

Continuing to pursue the same model for water development—the business-as-usual scenario—is unlikely to yield the requisite magnitude of results. This is so for three reasons.

First, the easy-to-achieve targets, essentially picking of the low-hanging fruits, have been achieved. UN-Water documents that this took place through greater investments into large, urban water/wastewater development projects (WHO 2010). This development approach has largely by-passed the much more challenging urban slums, peri-urban areas, and large swathes of rural

areas; each of these groups is very difficult to serve due to differing social, economic, geographical, and infrastructural problems. United Nations reports have accordingly documented huge disparities in provisioning of services between urban and rural areas (WHO and UNICEF 2014). Creating more large-scale water and wastewater treatment systems and large networks of water supply and sewage lines is unlikely to solve the problems in communities that are either too densely packed to allow installation of infrastructure or are too widely spread or too remote to create efficient infrastructure.

Second, governments and international donors are not in a position to mobilize the magnitude of capital that is needed. There is a growing sense that international aid alone cannot solve the water problems, and hence, national and sub-national governments must take on a greater burden of the implementation costs (Bigas *et al* 2012). At the same time, these same governments are being asked to respond to a multitude of social and economic challenges. As a consequence, the short-term political expediencies ostensibly trump long-term, strategic investments into water and sanitation services. The biennial high-level gatherings of economic and finance ministers from the least developed countries to discuss sanitation and water provisioning seem to represent, at a minimum, a greater recognition of the problem and its links to other development challenges (SWA 2014).

Third, the indigenous human resources and relevant expertise that are needed simply do not exist at the scale needed. The gaps in human capacity have not even been adequately estimated, but anecdotal evidence suggests that even when socially appropriate and economically feasible technological solution exist, their implementation is not possible due to the gaps in the human capital. Some researchers have suggested that scaling up of water and sanitation solutions—expanding to national or regional scale what is demonstrated successfully at village or community level—does not happen due to the human capacity gap (UNDP 2006).

The good news is that we are now at a phase in the international development dialogue in which each of these three reasons can be effectively addressed—through innovative solutions, new capital from the private sector, and capacity enhancement by the international development community. This opportunity has to do, in part, with the ongoing debate to formulate the future development agenda as the Millennium Development Goals reach 'maturity' in 2015 and a new set of Sustainable Development Goals (SDGs) take their place. The formal proposal for SDGs developed by the Open Working Group of the United Nations General Assembly, which explicitly listed targets for ensuring availability and sustainable management of water and sanitation for all, creates ample room for optimism.

Opportunities That Did Not Exist Before

The post-2015 international development debate has opened up options and opportunities that did not exist before. There are three such opportunities that are particularly relevant.

First, the private sector and the world business community have been invited to the table and have been actively engaged in the formulation of future goals and targets. One may argue that this engagement is driven purely by vested self-interest. While that is factually correct, after all each business has to be a profit-making enterprise to survive, the interest and engagement of the private sector is also driven by the reality that scarcity of resources, economic decrepitude, and social unrest are not good for business. In the water domain, many of the big international corporations have realized that unfettered exploitation of water resources with no re-routing of benefits to respective communities is not only a public relations disaster but also bad for business. Many of these corporations have thus channeled significant resources for working on building social capital, through community development activities, in tandem with the financial capital.

Second, the notion of 'green economy', new as it is, seems to be taking hold in the aftermath of the Rio+20 conference (UN-Water 2011). It offers a new perspective for the well-established sustainable development agenda, as originally defined in the Agenda 21 (UNCED 1992). The notion that creation of jobs, business entrepreneurship, and economic activity are integrally linked to sustainable development will likely help boost political uptake of the green economy concepts. The same concepts, and the underlying political benefits, also extend to the water sector and can be used to argue for greater level of investments at the national level.

Third, numerous options for crowd-sourcing and crowd-funding have emerged in the last few years, thanks to proliferation of online tools that facilitate this approach, which allow an average citizen to be actively involved in designing, funding, and implementing schemes for delivery of water and sanitation services. Although most of the examples of crowd-funding to date have focused on rural and community-level deployment of projects, there is no limitation of scale in applying these same solutions at national and international levels.

These new opportunities create the enabling environment for the private sector to be engaged in the field of activities related to the water challenges discussed earlier. However, one also needs to ask the question: why should it be engaged in this manner at all? The answer lies partly in the earlier discussion on why business-as-usual approaches have failed. Effective and well-managed engagement of the business community can overcome most of the shortcomings of water-related development paradigm. First, and most importantly, the private sector can put some much-needed financial capital on the table. It is clear that neither national governments nor the international donor community can mobilize the capital needed, which according to some early estimates could range between $1 and $2 trillion a year (UNU and UNOSD 2013). Second, by default, the private sector can lay claims to better business management in comparison to any public-sector enterprise (Batley 1996). A sharp eye on the bottom-line can ensure that water management and service delivery operations are run efficiently and economically. Third, the technical and technological know-how for almost all solutions and applications exists with the private domain. Fourth, the private sector has ample experience in scaling up and operational growth, often by orders of magnitude; in contrast, the public sector enterprises almost never encounter logarithmic growth and could in part be the limiting factor in why success stories are not readily scaled up from local to national levels.

However, such engagement also comes with its own cautionary tales and pitfalls.

Stumbling Blocks in Engaging the Private Sector

The largest stumbling block in the engagement of private sector in water-related enterprises is that of public perception. There is a public perception that water is a 'public good' and privatizing its provisioning or its commodification (Barlow 2001) violates the public trust. While this argument ignores the fact that pumping, treatment, provisioning, maintenance, billing, and overall management are essentially business activities and need to be run on a cost-recovery model, its emotional appeal to the general public is considerable. Demonizing the private sector, and often those who associate with them, is not difficult given the general distrust in public and a wide range of online and social media outlets available. Numerous case studies of failure of the private sector to efficiently manage and deliver water services further exacerbate the situation. That essentially means that any commercial enterprise that engages in the water sector becomes a ready target for activist groups.

Another major stumbling block is to get the right mix of public- and private-sector engagement. The relationship between the two sectors of economy can range from the traditional command-and-control approach to a more market-oriented relationship (Paoletto and Termorshuizen 2003). The latter approach has been shown to be more successful, but achieving the balance between regulation and facilitation remains more of an art form than well-established science.

Lack of enabling environment, including over-regulation of the sector and absence of security for capital investments, is another major turnoff for the private sector. One key indicator of this problem is the magnitude of venture capital that becomes available for the water sector. For example, in North America the magnitude of venture capital vested in the water "sector" dwarfs that available to other segments of the economy by at least an order of magnitude. In the absence of a major political push, the pubic regulators are quite reluctant to offer leeway or even due consideration to innovations in business models, technology applications, or service delivery. A positive example was set by the province of Ontario, which in 2010 approved the "Water Opportunities Act" as a means of better engaging the private sector and creating new economic activities. A number of ensuing actions, including a Water Technologies Acceleration Project (or WaterTAP), have demonstrated that creating the right incentives can yield positive results in both economic and political sense.

Gaps in knowledge and scientific evidence play a relatively minor role in achieving gainful engagement of the private sector. Nonetheless, the most obvious gap is in not having successful business models for provisioning of water (and sanitation) services to the so-called "base of the pyramid" segment of the society; a euphemism for poorest of the poor. Even if positive public perception and effective public oversight are achieved, the question of how to turn a profit in the short term remains unanswered. We need a large number of pilot studies that test different business models in a variety of settings to generate the requisite empirical evidence. Alternatively, a more systems-based approach may help us better understand the key elements for success of a business model. Application of untested, but potentially revolutionary water treatment technologies, also presents a gap in our knowledge—one that can typically only be filled through on-the-ground implementation of pilot projects. Finally, we do not fully understand the risks involved in not effectively managing water resources and delivering water services; meaningful, evidence-based risk analysis must provide usable baseline information.

Conclusion: Incentivizing the Key Players in the Private Sector

Even a brief, superficial analysis, such as that presented in this paper, can lead to the conclusion that engaging the private sector in delivering solutions to the global water crisis makes eminent sense. Nevertheless, significant stumbling blocks remain, which must be addressed before this engagement can take place successfully. Some of these hurdles should be directly addressed as the post-2015 development agenda takes shape and its operational elements are better understood. The approaches for engaging the private sector in social development would extend far beyond the water sector and into other domains including *inter alia* health, food, energy, and education. A direct involvement of the business community in this discovery process can obviously be of mutual benefit.

The key point is how to create positive incentives and effective regulations. The answers would vary from country to country, and no magic bullet exists at the moment that would work everywhere. Broad stakeholder consultations that involve civil society organizations and community representatives as well as public- and private-sector representatives are essential to discovering business and delivery models that operate in particular social and economic settings. The international development community, including the United Nations system, has a major role to play in facilitating these dialogues and enhancing local and national capacities to implement selected solutions.

References

Barlow, M., 2001. "Blue Gold: The Global Water Crisis and the Commodification of the World's Water Supply." International Forum on Globalization. San Francisco.

Batley, R., 1996. "Public-Private Relationships and Performance in Service Provision," *Urban Studies* 33: 723.

Bigas, H., Morris, T., Sandford, B., Adeel, Z. eds. 2012. "The Global Water Crisis: Addressing an Urgent Security Issue." Papers for the InterAction Council, 2011–2012. Hamilton, Canada: United Nations University Institute for Water, Environment and Health (UNU-INWEH).

National Intelligence Council (NIC), 2012. *Global Water Security—Intelligence Community Assessment—ICA 2012-08, 2*. Washington DC: Office of the Director of National Intelligence.

Paoletto, G. and Termorshuizen, C. 2003. "Chemical Governance in East Asia." In *East Asian Experience in Environmental Governance: Response in a Rapidly Developing Region*, ed. Z. Adeel. Tokyo, Japan: UNU Press.

SWA, 2014. "Partners in Action, The Voice of the SWA Partnership", Issue 3, June 2014. New York: Sanitation and Water for All (SWA).

UNCED, 1992. *Agenda 21*, United Nations Conference on Environment & Development (June 3–14, 1992). New York: United Nations.

UNDP, 2006. Human Development Report 2006—Beyond scarcity: Power, poverty and the Global Water Crisis. New York: United Nations Development Programme (UNDP).

UNEP, 2012. The UN-Water Status Report on the Application of Integrated Approaches to Water Resources Management. Nairobi, Kenya: United Nations Environment Programme.

UNU and UNOSD, 2013. "Water for Sustainability: Framing Water within the Post-2015 Development Agenda." United Nations University Institute for Water, Environment

and Health (UNU-INWEH), UN Office of Sustainable Development (UNOSD), and Stockholm Environment Institute (SEI). Hamilton, Canada: UNU-INWEH.

UN-Water, 2011. "Water in a Green Economy," in *A Statement by UN-Water for the UN Conference on Sustainable Development 2012* (Rio+20 Summit), http://www.uncsd2012. org/content/documents/303UN-Water%20 Rio20%20Statement%201%20NOV.2011. pdf.

UN-Water, 2013. "Water Security & the Global Water Agenda." A UN-Water Analytical Brief. Hamilton, Canada: United Nations University Institute for Water, Environment and Health (UNU-INWEH).

WHO, 2010. "UN-Water Global Annual Assessment of Sanitation and Drinking-Water (GLAAS) 2010: Targeting Resources for Better Results." Geneva, Switzerland: World Health Organization.

WHO and UNICEF, 2014. "Progress on Sanitation and Drinking-Water—2014 Update." Geneva, Switzerland: World Health Organization and United Nations Children's Fund,

WWAP (United Nations World Water Assessment Programme), 2012. The United Nations World Water Development Report 4: Managing Water under Uncertainty and Risk. Paris, France: UNESCO.

WWAP (United Nations World Water Assessment Programme), 2014. "The United Nations World Water Development Report 2014: Water and Energy." Paris, France: UNESCO.

Water on The African Continent and Its Use and Benefits for Development: A Vision for Angola

Carmen dos Santos[A]

In this opinion editorial, New Water Policy and Practice International Advisory Board member Dr Carmen dos Santos (Director of the Department of Biology in the University of Agostinho Neto, Angola) discusses the challenges and the recent development of the institutional framework for water management in Angola in the context of the Southern African Development Community (SADC), with particular focus on the recognition of the strategic importance of water for the regional economic integration, and on the management of shared water resources.

Keywords: institutional framework; water governance; development; shared water resources; SADC

Introduction

Talking about water in Africa is to speak of a precious commodity, sought by all, but to which few have access. A large part of the population on the continent lives with few resources that do not allow access to this so precious good. In general, the urban population which accesses piped water for consumption is less than the population of the periphery, where access to water is rarely a structured and available service. Access to water for the rural population also poses a major problem, without losing sight of the broader interests of rural communities. This is not only to provide water services to the population for human consumption and economic activities, but also to ensure that water is available for all actors in a holistic context and environmentally healthy.

In recent years, the African continent has been evolving in terms of human and social development, along with the economic development of most countries. It is expected that Africa can manage its potential in water and that this is possible through the adoption of new policies that aim to incorporate research results and scientific knowledge.

Angola is going through a phase of political, technical and, economic development with an integrated strategy that is allowing the national water program to be deployed to all while diversifying the multidisciplinary studies for the different river basins. There is still a lot to do in order to ensure that all actors have access to water, secured and distributed through accepted international best practices. Public policies regarding the services to the population, and in particular water, seem to be increasingly consistent and addressed to the more needy populations.

[A] Carmen dos Santos, Agostinho Neto University, Angola

Water in Angola

The policies on water have evolved greatly over the decades and Angola is no exception. At the national and regional context, there is a greater concern with providing the population with services that allow their increasing well-being. Nowadays in Africa, and in Angola, there is a growing concern for the environment because of greater awareness and ability to execute the Government policy. The Angola Government performs various framework programs aimed at combating poverty, with particular focus on access to water. These objectives align with food security, guarantee of access to school, literacy, and access to health and the professionalization of the youth to obtain their first job.

These guarantees were stablished in the Angolan Constitution and are regulated by sectoral laws that ensure, in general, the implementation of the national and provincial development plans that guide the process of structural changes. In the regional context, Angola integrates into the economic region corresponding to the Southern African Development Community (SADC), whose programs and projects on water have been of paramount importance. Recognition of the strategic importance of water for the regional economic integration and the management of shared water resources is also an objective of the SADC.

Water is, to all living beings, a priority need and access to it can constitute a major difficulty leading to energy consumption, expenditure of time and money for the general population. Government policies are often based on the existing natural water availability in the country, with higher incidence of surface waters but often underutilized groundwater resources. Africa is starting to promote and adopt policies for a better knowledge of their natural resources. This is happening through both greater diversification of economies and greater demand of society over their governments.

In Africa, the water issues are still associated with sociological, cultural, and historical aspects that interconnect with other facts, such as the low development of rural populations and, in particular, the precarious living conditions of women. On the other hand, there are major environmental conflicts coming from both the development of dams along rivers and disputes over arable land and access to the river banks.

Angolan scientists are increasingly called upon to provide more accurate knowledge regarding surface and underground water resources. Overall, 77 drainage basins were defined, of which 70 are fully within the borders of Angola.

The Constitution of the Republic of Angola safeguards water as a public good and belonging to the State. The regulation of the water sector, Water Law, Act No, 6 "/02 of June 21, provides guidance through Article 5: the waters while natural resources are owned by the State and are part of the public water domain. This is reinforced in Article 20 where the uses are classified into common and private. Note that for private uses the law allows a permit or concession, but always preventing that the private interests overlap the needs of water supply to the population.

The present Government developed a strategic plan for the water sector expected to be in force in the period 2004–2016, and adjusted after the elections of 2012, with specific objectives for the rapid structuring of services to the population. This strategic plan is integrated with the National Development Strategy 2012–2025 and in the National Development Plan 2012–2025. Public policies defined by the Angolan authorities for the water sector provide clear priority guidance: water supply for all.

In short, the strategy involves the creation of offices for the administration of river basins at the level of the whole country. This guidance aims to support accelerated development of the use of catchment areas for water supply and hydroelectric capacity. The strategy emphasises that these governance offices will be established in 47 basins that must be managed appropriately and environmentally sustainably to meet the needs of the population.

The need to ensure power supply to the populations also creates a great deal of pressure to the water resource. We see today, even if not very rooted in the basic concepts of development, government policies orienting to the development of mini hydropower as opposed to large dams, within an ecological perspective of the management of basins. According to the government, all projects for water supply need to be regulated and provided through a system of permits and concessions to be awarded to private entities. In other words, the privatization of water becomes a reality in Angola, although it has been stated that the concessions will be public–private partnerships. "Private entities can join now, because the law already provides the approval of the regulatory and legal support to private partners," said the Secretary of State of the Waters.

In general, Angolan cities are at an initial stage of implementation of their Master Plans where service demands through household connections are previewed. The need for real implementation of the municipal management of services is included in one of the recommendations of National Forum on the Water for All program. There the need for adoption of mechanisms for coordination between Provincial Directorates for Energy and Water along with the municipal administrations as the entities that are responsible for the implementation of the integrated municipal rural development and combating poverty programs (cit. ANGOP 3/3/2014) was established.

In the urban areas, the users of municipal services of water are the most interventionist and clamoring for this much-yearned resource, making access to water in these areas more structured through the presence of public water companies in each city. In rural areas, the situation is far more complicated for average users who need to walk dozens of kilometers to get water from rivers. The rural population is poorer and the literacy rate is too low, implying that there is almost non-existent knowledge and ability of the citizens to demand the fulfillment of their constitutional rights.

This fact is associated with socio-cultural aspects that generally allow common profiling of all the peoples in the different Angolan regions. It is very important to mention that a woman in this society has an important role within the family, as a matriarch in the practical sense of the word, and therefore she runs between various tasks that include providing the subsistence farming, gathering of firewood, water, caring for the children, and performing all domestic services.

But access to water is not just an issue for humans, it is also important for wildlife and water animals, as members of the environment that we all share. Environmental awareness should be a part of the substantive agendas of all government policies since the environment and the crosscutting issues are linked. The policy on quality and use of water has yet to regulate certain uses and, above all, to provide the ecological characterization of all watersheds, enabling water-use scenarios for each environmental component.

In the context of these policy measures that are to be implemented in the short and medium term, all the guidelines of sustainable development provided by integrated planning will aim at implementing more rational use and prevent pollution of water bodies by misuse. Between the national and international basins, it will also advise the creation of management networks in order to connect the local, regional, and national contexts.

This integrated vision may provide greater knowledge, designed for policy makers, about the demand of water, and how to share development benefits. This approach aligns with that put forward by Álvaro Pereira (2008) which states that programs and governance priorities face mainly structural and conjuncture constraints such as: the level of availability of water resources; cultural, socio-political, economic, and environmental mindset; governance variables within dominant geopolitics and economic order; and finally the ideological wrangling and administrative provisions.

In Angola, in terms of environment, the guiding legal framework contributes several principles, namely the environment framework law 5/98 which establishes the general environmental paradigms and different complementary executive decrees and regulations which are in full compliance. It should be noted that the law on spatial planning and urbanism 03/04 and the land law 9/04a) are also regulatory instruments that guide these activities.

In the context of recognition of the role of organized civil society through the Associations Law for Environmental Protection 03/06, which supports the environment and organizes public participation, the transversal aspect of environmental issues is an open door for defining new paradigms on water and the sharing of this resource.

In the current state of development, Angola still maintains a good level of availability of water resources, but despite the considerable results already achieved, still faces structural level constraints and the need to overcome barriers imposed by different situations from the moment of conception until the time of execution of the programs.

The future development of water policies in each country also includes recognition and accomplishment of the various UN guidelines, namely (UNITED NATIONS, 2003):

i) Protection of ecosystems and guarantee of its integrity through sustainable management of water resources;
ii) Sharing water resources promoting peaceful cooperation between different uses of water and between States, through approaches such as sustainable management of watersheds;
iii) Risk management to provide security from a range of water-related hazards that may occur;
iv) Valuing water to be managed on the basis of their different values (economic, social, environmental, cultural) and move towards the fair price of water and recover the costs of the provision of services, taking account of the social capital and the needs of the poor and most vulnerable; and
v) Regulate water wisely, involving the public and the interests of all stakeholders.

In the political perspective of the development, a framework is created where access to water is a process that must be developed and managed through efficient and effective public policies.

In Angola, the goals of the Water for All program should be evaluated to identify alternatives that allow the speeding up of the process of improving access to water for urban, peri-urban, and rural population.

New goals must be generated through a holistic approach to access to water. New generations of players/users are required to add value to the process, enriching it, submitting specific global counterproposals or underscoring a new innovative and responsive vision. The idea is always to achieve high levels of social welfare for the population through sustainable development.

What We Expect And How To Contribute To Problem Solving?

We will use this platform to publish scientific approaches on the study and dissemination of new policies on water. The new approach should be seen as a way to galvanize the future followers of this new Journal whose fundamental objective is to provide a new vision, a new look at the public policies on water worldwide, renovating the old assumptions that have brought disappointing results out of the current global context. We encourage all of you to develop innovative ideas within the framework of good environmental practices.

References

African Development Bank. 2014. "Gender, Poverty and Environmental Indicators of African Countries." Published by the: Economic and Social Statistics Division Statistics Department. African Development Bank. Temporary Relocation Agency (TRA), Tunis: ADB

ANGOP, 2014. (Secretary of State of Energy and Water on 3/03/2014).

MacDonald M., Bonsor, H. C., Dochartaigh, O. and Taylor, R. G. 2012: "Quantitative maps of groundwater resources in Africa." Environ. Res. Lett. 7.

ONU. 2003. "Water for People, Water for Life." UN World Water Development Report (WWDR) (co-published with Berghahn Books, UK).

Pereira, Á. 2011. "Água em Angola: a insustentável fraqueza do sistema institucional." Revista Angolana de Sociologia [Online], 8, posted online on 29 July 2013. http://ras.revues.org/519; DOI: 10.4000/ras.519 (accessed August 2, 2014).

Governo de Angola. 2010. "Constituição da República." Angola.

Governo de Angola. Lei das Águas (*Water Act*), Lei nº 6/02 de 21 de Junho. Governo de Angola.

SWECO GRONER. 2005. "Avaliação Rápida dos Recursos Hídricos de Angola." Final Report, Luanda: Direcção Nacional de Águas.

Integrated Water Resources Management from Rhetoric to Practice

Shahbaz Khan[A]

In this opinion editorial, New Water Policy and Practice International Advisory Board member Prof Shahbaz Khan (Deputy Director UNESCO Regional Science Bureau for Asia and the Pacific, Jakarta, Indonesia) presents some ideas on how to move Integrated Water Resource Management (IWRM) from rhetoric to action. Reflecting on IWRM, and on some of the difficulties experienced in its implementation, Shahbaz discusses some international initiatives which can strengthen IWRM in practice, identifies obstacles to operationalising IWRM principles and proposes some opportunities to facilitate IWRM implementation at the river basin level.

Keywords: *IWRM; integrated approaches; sustainable development; river basin management; IHP-HELP*

I - Introduction

Water resources management played a crucial role in the fate of earliest civilizations and is continuing to remain a significant link to sustainable development (Khan et al. 2006). New global change drivers are intensifying water, food security and energy inter-linkages thereby creating unprecedented positive and negative impacts which need to be managed together (UNESCO 2014). Water also remains a critical resource for human health, development and prosperity in the post-2015 development agenda. The challenges posing water management, including catchment care, provision of regular water supplies, improved sanitation, increased food production, and prevention and reduction of water related disasters, need to be addressed in ways that restore and pro-

tect environmental and ecosystem quality. This is possible only if there is a common recognition that the sustainable future of water management should begin with the conservation and restoration of landscapes and underlying aquifers which are often strongly connected to rivers.

Changing complex socio-economic interactions with water, carbon-energy, food production and climate cycles are putting new pressures on land, water, and associated ecosystems. This requires application of trans-disciplinary integrated participatory approaches to restore, enhance and protect sustainability of land and water systems. In recent years there is an increasing emphasis on the integrated approaches to land and water management e.g. the Integrated Water Resources Management (IWRM), Integrated Lakes Management, Landscape Approach. Such approaches are necessary

[A] UNESCO Regional Science Bureau for Asia and the Pacific, Jakarta, Indonesia

to open the paradigm lock between research community and policy-makers who are struggling to manage complex interactions between biological diversity, climate change, land use change, and freshwater use limits and constraints. These represent four of the nine boundaries of the Earth System processes recommended not to be crossed to avoid unacceptable environmental change to humanity (Rockström et al. 2009). While integrated approaches are considered as panacea by their ardent advocates there is a growing criticism by others due to individual and institutional capacity limitations, funding and policy constraints. These constraints differ for individual water sectors, therefore the enabling environment needs to cater for differing land and water management challenges at the river basin level.

Given the fundamental role of water for sustainable development, the need for an integrated approach to water resources management and the role of ecosystems for achieving a water-secure world, has also been reaffirmed in the outcome document of the Rio+20 United Nations Conference on Sustainable Development, entitled "The Future We Want" (paragraphs 119-124) adopted by the UN General Assembly through resolution A/RES/66/288. The pivotal role of water for socio-economic development and for maintaining healthy ecosystems is more and more threatened due to intensifying pressures on water resources. The population growth, the demands on hydro energy, the higher rates of urbanization all over the world, demands for drinking and industrial water, the degradation of ecosystems and environment, and the impacts of climate change are among the key challenges to be addressed.

This paper focuses on integrated water resources management in the context of the overall concept, international initiatives, and proposals to promote it while overcoming identified obstacles.

II - The IWRM Rhetoric

While devising Sustainable Development Agenda, the implementation of IWRM at the river basin level can offer pathways to implement water solutions. It is based on generic principles, approaches and guidelines formulated at the International Conference on Water and the Environment, Dublin in 1992, aiming to promote changes in concepts and practices which are considered fundamental to improve the management of water resources. These principles were further reaffirmed in the outcome document of the Rio+20 United Nations Conference on Sustainable Development, (Paragraphs 119-124) which reaffirms that IWRM does not need to be seen as "*a dogma*" or "*a de facto solution*." In fact one of the main advantages of IWRM approach is its flexibility and adaptability according to the circumstances. Maybe the generic principles remain the same but their interpretation and practical implementation needs local customisation. It is important to take into account the hydrological specificities of a given river basin as well as socio-economic setting with cultural sensitivity. For more efficiency IWRM has to be implemented through local leadership and acceptance of all stakeholders.

It is recognised that much remains to be done in terms of designing, financing and implementing an integrated approach to water resource management as stated in the 2002 Johannesburg Plan of Implementation.

While acknowledging the goals and importance of IWRM concept and approach, it is also been argued by some workers that IWRM "*has basically been turned into a dogma [that has become] a de facto solution for many water issues*" making IWRM look as a lofty initiative, *not appropriate for all times and circumstances*, hard not only to convert its noble concepts into practice, but "*sometimes results in negative outcomes, policy failure and has shut out alternative thinking*" (Giordano 2012).

It has been argued by several water experts and specialised institutions that water problems of the world are neither homogenous, nor constant or consistent over time and often vary very considerably from one region to another. Solutions to water problems depend not only on the physical availability of water, but also on many other socio-economic and policy factors. Some of these factors include management processes, institutional competence and capacities, prevailing socio-political conditions that dictate water planning, technical processes and practices, appropriateness and implementation statuses of the existing legal frameworks, availability of investment funds, social and environmental conditions of the countries concerned, stakeholder perceptions, modes of governance including issues like political interferences, transparency, corruption, the educational and development conditions, and status, quality and relevance of research that are being conducted on the national, sub-national and local water problems (Biswas 2004).

Much criticism of the design and implementation of IWRM policies and programmes is not about the lack of technical solutions, but on the poor institutional organisation and/or to the insufficient legislation, the enforcement of water laws and regulations. Institutional and legal frameworks are key elements of IWRM, however, in many developing countries, water institutions remain too weak or too young to adequately carry out IWRM and need therefore to be strengthened in the domains of policy, research and monitoring. International programs have to play a significant role for the establishment and/or strengthening of IWRM institutions, as a fundamental element for a water secure future for all.

III - Some International Initiatives To Strengthen Implementation Of IWRM

UNESCO Natural Sciences Sector through the International Hydrological Program (IHP) and the Man and Biosphere (MAB) Program has been promoting trans-disciplinary approaches for land and water management for better environmental, social and economic outcomes. The IHP Ecohydrology approach (Zalewski, Janauer, and Jolánkai 1997; Zalewski 2002; Zalewski and Wagner 2004) involves the development of tools that integrate basin-wide human activities with hydrological cycle in order to sustain, improve and restore ecological functions and services in river basins and coastal zones as a basis to support positive socioeconomic development. The IHP Hydrology for Environment, Life and Policy (HELP) initiative in river basins aims to deliver social, economic and environmental benefits to stakeholders through research towards the sustainable use of water for human and environmental purposes. This is accomplished by linking hydrological science with improved integrated catchment management from plant to river basin levels. The ecosystem services approach can help in improving the complex relationships between hydrological processes, water resources management, ecology, socio-economic and policy-making in HELP river basins. Through IHP-HELP the ecosystem services approach has been promoted among scientist, policy-makers and water practitioners as new ways of solving the complex upstream and downstream stakeholder issues and for bringing consensus among competing water users. The Ecohydrology initiative aims to enhance ecosystem services through dual regulation of flow and biota in fresh water and estuarine environments. The ecosystem services approach also offers an adaptable tool to river basin managers who are struggling to implement IWRM for better balance between consumptive and environmental uses of water. In this regard UNESCO IWRM Guidelines at the River Basin level can benefit from practical case studies which should be invited to be published in this new journal. This will also increase IWRM database of keys for success to implement integrated solutions through an upward spiral approach which maximises the triple bottom line. Experiences gained through previous UNESCO IHP phases have shown that freshwater availability will become a major concern if no immediate actions are taken to restore and enhance the associated ecosystems.

UNESCO has been collaborating with the Network of Asian River Basin Organizations (NARBO) to develop the IWRM Guidelines at the River Basin Level to address practical implementation of IWRM. IWRM is essentially a user-friendly and cooperative approach that invites each sector to meaningfully participate and cooperate in the implementation, with a practical roadmap so as to contribute to achieving both private and public benefits in a sustainable manner (UNESCO 2009; Nakajo 2010). As part of these guidelines UNESCO's has been promoting a Spiral Model (Figure 1) for integrated water resources management which involves several repetitions of four-stage (recognizing, conceptualizing, coordinating and implementing) water resources development in a river basin. It recognizes that management principles and objectives evolve over time as new demands and needs

emerge, and innovative solutions are added at each stage. The development stage changes when recognition of the need for change arises among stakeholders. Moving up the spiral is a participatory management process, which requires reaching agreements with stakeholders and building new consensus. While the 'spiral model' is a convenient graphical conceptualization of the iterative, evolutionary, and adaptive management process, for adjusting to new needs, circumstances, and societal goals there is a need for a clear understanding of the approach by all stakeholders. The institutional and policy frameworks need to be adapted over time to the evolutionary nature of the spiral model for progressive positive changes in water resources development and management.

IV - Synthesis of obstacles to the implementation of IWRM principles

Some obstacles to wider adoption of IWRM practice identified through international initiatives are described below.

Lack of knowledge of water balance and its interactions with water quality: the impacts of biophysical changes in catchment hydrology on water quantity and quality remain less well defined (Khan, Rana and Hanjra 2010a). A better understanding of the fate of contaminants and pathogens passing through the water cycle remains only desirable knowledge in most situations.

Missing understanding of the Ecohydrology of river basins: there is still poor ecohydrology knowledge at the landscape level. Specialists in IWRM need to improve understanding of water-landscape level management of the environment, taking full account of the interactions among ecosystems and their dependent habitats. Inter-linkages between agro-biodiversity and land and

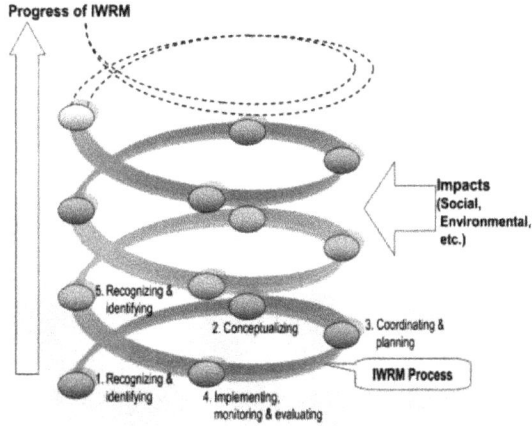

Figure 1: IWRM Spiral and Process (IWRM Guidelines at River Basin Level—Part 1: Principles)

Figure 2: Need to bridge gaps between sectoral approaches for management (Source Khan, Savenije and Demuth 2010b)

water management remain vaguely defined in catchment planning processes. Changes in land cover on the reduction/increase in the recharge potential for aquifers as well as changes in biogeochemistry with water cycle alteration may generate unavoidable environmental problems due to the physical and chemical landscape manipulations they involve. In the worst case, human interventions can affect ecosystems to the point where they are unable to deliver ecosystem services, such as fresh water, productive soils or maintenance of valuable biodiversity — with direct consequences for livelihoods, vulnerability and security.

Lack of quantification of surface-groundwater interactions and their policy implications: Lack of scientific knowledge about river–aquifer interactions due to disciplinary and institutional divides remain a major hurdle in integrated management of surface and groundwater. Successful IWRM requires closer attention to stocks and flows of surface and aquifer water balances, equity in the use of groundwater resources, and adaptation measures to climate variability, groundwater quality, groundwater protection, groundwater-dependent ecosystems and urban groundwater management.

Sectoral financing and management approaches: Water financing and management is divided by agriculture, urban, industry sectors (Figure-2) has failed to recognize the true environmental costs and benefits of water investments across the river basins and aquifer systems. For example financing of most irrigation dams fail to link with upper basin forest replanting, soil conservation, and downstream environmental consequences. Financing sustainable management of hydrological units remains hindered by a lack of conformance with political/administrative boundaries.

Absence of real stakeholders from water solutions: Water problems are becoming more complex and acute all over the world due to pollution of resources and inequity of supplies between different stakeholders. The water problems are becoming super wicked (Khan et al. 2008) due to growing population, widespread mismanagement and corruption, degraded sources by unidentified pollutants with unpredictable effects. Older cities are facing critical challenges, such as deteriorating infrastructure, a degrading environment and an inability to participate due to lack of rigid regulators will and conflicting investment demands by transport and energy sectors.

Lack of champions for effective cross-disciplinary dialogue: There is a general lack of integrators, people who can lead cross-disciplinary dialogues for solving IWRM problems by bringing together biophysical scientists, sociologists, policy makers and implementers through common understanding. People responsible for solving existing water problems are often leading to new problems rather than facilitating solutions. The traditional water specialists are afraid of implementing IWRM due to longer time period needed for achieving community consensus. Lack of rewards and complex expert semantics are a major hurdle to nurture champions of cross-disciplinary solutions.

V - Some propositions to facilitate IWRM practice at the river basin level

Although IWRM concept is now a well-accepted approach worldwide, the main challenge remains on how positive social, economic and environmental change can be made in real catchments while working with real people. *There is a need to listen to critics of IWRM concept and learn from differing opinions e.g. there is a valid argument that good water management is not only more integration but also the promotion of what can be called a 3 C approach which means more "Collaboration," "Co-operation," and "Coordination"* (Biswas 2004). One of the main hurdles to the implementation of IWRM is a general lack of capacity of trans-disciplinary champions from least developed countries and disconnected communities lacking voices in decision-making processes. There is a greater need for focus on micro financing of community projects, community empowerment through long-term water resource planning, purpose driven engagement for pragmatically working with all stakeholders through appropriate economic instruments and targets (Heath 2010).

At this critical juncture of time when the world is working on the post-2015 development agenda, IWRM should no longer be treated as a new paradigm but as an interactive process for empowering communities to alleviate poverty and to achieve basic minimum standards of water and sanitation for human dignity. IWRM should not be treated as the only prescription for sustainability, but as a practical framework for addressing water management challenges through capacity building and community engagement.

In the process of facilitating the implementation of IWRM at the river basin level, the aspirations of real stakeholders need to be aligned with priorities of the international programs, efforts of research and training institutions, networking of river basin organizations. Policy makers and decision takers need to be engaged in a practical facilitating manner. The following processes can lead to better water futures.

• *Establish IWRM centres of excellence*: The research and training on IWRM needs to be promoted through formal system of learning which bring together researchers, managers and policy makers. The establishment of IWRM centres under international initiatives, such as the UNESCO Category 2 institutes/centres could help to develop a network across the world working on sectoral issues, but complementary for the success of IWRM in solving keys water challenges of the future.

• *Take advantage of trans-disciplinary initiatives*: Programmes that are designed to incorporate relevant policy and scientific issues through cross-cutting approaches on water management should be promoted for the implementation of IWRM. In this regard UNESCO-IHP programmes such as Hydrology for Environment, Live and Policy (HELP) and Ecohydrology could play a significant role if implemented in collaboration with river basin organisations.

• *Establishment of IWRM for delivery Sustainable Development Goals (SDG) demonstrations sites*: The IWRM demonstration sites/projects will play a significant role to convince critics that this concept as a relevance to solve SDG-related problems in real river basin. Such an approach is not only necessary for testing and dissemination of the IWRM concept, but also for advancement of its worldwide implementation, and substantial contribution to sustainable use and management of water resources for sustainable societies.

• *Promote IWRM cooperation in transboundary rivers and aquifers*: There is urgent need to enhance transboundary cooperation for peace especially on water management-related issues involving countries with shared river basins and aquifers. This calls for greater cooperation across scientific, social, economic and environmental boundaries through appropriate international treaties. Co-operation on IWRM could be structured around south-south and/or triangular cooperation in coordination with professional associations, river organizations, NGOs and intergovernmental institutions.

VI - Epilogue

The IWRM concept has emerged as a plausible way to bring a multi-stakeholders approach to worsening water security from floods, droughts and bad water quality. However, the need for continued stakeholder empowerment in managing water across sectors seems to be a challenge for traditional managers to adopt this approach across all uses of water. A number of institutional, policy and capacity issues have emerged with growing disbelief of managers whether IWRM can be applied at a practical level. While acknowledging complex implementation challenges, IWRM remains as a key tool for conceptualising, planning, implementing and managing wicked water problems through broader consensus of competing stakeholders. UNESCO, wider UN and International NGOs are working on a number of initiatives to help stakeholders achieve water security through local and global actions using IWRM at the river basin level, however their efforts must have local ownership by river basin stakeholders. Despite its all shortcoming IWRM continues to offer integration across sectors, programmes and groups of society over time. There is a need to design and implement individual activities to facilitate longer term positive societal change in river basins. IWRM is not a silver bullet to solve all water problems but a coordinated process to be followed step by step and level after level over time in an evolutionary and adaptive manner.

References

Biswas, A.K. 2004. "Integrated Water Resources Management: A Reassessment." *Water International* 29 (2): 248-256.

Giordano, M. 2012. "Non-Integrated Water Resources Management." IWMI. November 2012. http://www.slideshare.net/IWMI_Media/nonintegrated-water-resources-management.

Heath, T. October 2010. "Pragmatic But Principled." Background Report on Integrated Water Resources Management, Cranfiled University. http://www.wsup.com/sharing/documents/IWRMBackgroundReport2010.pdf.

Khan, S. 2008. "Turning Concepts Into Community Driven Catchment Water Management Solutions." Foreword to the *special HELP edition Water* SA 34 (4): 429-431.

Khan, S., Rana, T. and Hanjra, M.A. 2010a. "A Whole-of-the-Catchment Water Accounting Framework to Facilitate Public–Private Investments: An Example from Australia." *Water Policy*. doi:10.2166/wp.2009.2027.

Khan, S., Rana, T., Yuanli, C. and Blackwell, J. 2006. "Can Irrigation be Sustainable?" *Agricultural Water Management* 80: 87-99.

Khan, S., Savenije, H. and Demuth, S. 2010b. "Tools for Analysing Hydrocomplexity and Solving Wicked Water Problems: A Synthesis." In *New Tools for Solving Wicked Water Problems*, eds. S. Khan, H. Savenije, S. Demuth, and P. Hubert. IAHS Publication 338, 145-158, Wallingford.

Nakajo, Y. 2010. "A Spiral Approach to IWRM: the IWRM Guidelines at River basin Level." In *Hydrocomplexiety: New Tools for Solving Wicked Water Problems*, eds. S. Khan, H. Savenije, S. Demuth, and P. Hubert. IAHS Publication 338, 145-158, Wallingford.

Rockström, J., Steffen, W., Noone, K., Persson, Å.,Chapin, III, E. Lambin, T.M. Lenton, M. Scheffer, C. Folke, H. Schellnhuber, B. Nykvist, C.A. De Wit, T. Hughes, S. van der Leeuw, H. Rodhe, S. Sörlin, P.K. Snyder, R. Costanza, U. Svedin, M. Falkenmark, L. Karlberg, R.W. Corell, V.J. Fabry, J. Hansen, B. Walker, D. Liverman, K. Richardson, P. Crutzen, and J. Foley. 2009. "Planetary Boundaries: Exploring the Safe Operating Space for Humanity." *Ecology and Society* 14 (2): 32. http://www.ecologyandsociety.org/vol14/iss2/art32/.

UNESCO. 2009. IWRM Guidelines at River Basin Level. Part 1: Principles, UNESCO-International Hydrological Programme (IHP), World Water Assessment Programme (WWAP), Network of Asian River Basin Organizations (NARBO).

UNESCO. 2014. "The World Water Development Report." Water and Energy. http://www.unwater.org/worldwaterday/world-water-development-report/en/.

Zalewski, M., ed. 2002. Guidelines for the Integrated Management of the Watershed. Phytotechnology and Ecohydrology. UNEP-DTIE-IETC, UNESCO-IHP, Freshwater Management series No. 5., Osaka, Siga, pp. 188.

Zalewski, M., G.A. Janauer, and G. Jolánkai. 1997. Ecohydrology. A new paradigm for the sustainable use of acquatic resources. Conceptual background, working hypothesis, rationale and scientific guidelines for the implementation of the IHP-V Projects 2.3/2.4. IHP-V Technical Documents in Hydrology No. 7, UNESCO, Paris, pp. 58.

Zalewski, M., and I. Wagner-Lotkowska., eds. 2004. Integrated Watershed Management—Ecohydrology & Phytotechnology—Manual. Osaka, Shiga: UNEP-DTIE-IETC, pp. 208.

Transboundary Wastewater Management Under Conditions of Inadequate Infrastructure and Political Complexity: the Need for Decentralized Approaches

Clive Lipchin[A]

In this opinion editorial, New Water Policy and Practice International Advisory Board member Dr Clive Lipchin (Director of the Center for Transboundary Water Management, Arava Institute for Environmental Studies, Israel) discusses the need for decentralized approaches to transboundary wastewater management under conditions of inadequate infrastructure and political complexity. Clive focuses on the need for local approaches, stressing the relevance of community level management approaches and, using the Israel-Palestine context in Gaza as a case example, calls for adaptation to socio-political challenges to find long-lasting solutions.

Keywords: *transboundary wastewater management; decentralized approaches; political complexity; stakeholder engagement; Israeli-Palestinian context*

Over 2.4 billion people worldwide lack sufficient wastewater treatment infrastructure, primarily due to the high costs of building and operating centralized wastewater treatment facilities, limited local expertise, and the lack of sound wastewater management policies (Massoud, Tarhini, and Nasr 2009). Poor wastewater management results in pollution of raw sewage to surface water and groundwater. In addition to severe environmental degradation, the public health impacts of water pollution are urgent and widespread across the globe; over two million people die each year due to contaminated water supplies (WHO 2012). To mitigate threats to the environment and public health, coordinated wastewater treatment is essential.

Decentralized, regional strategies or national regulatory approaches represent wastewater management options that may be appropriate for different sociopolitical contexts. In recent years, an integrated water resources management (IWRM) approach has dominated the discourse. IWRM is a set of principles that integrates social and hydrological factors to promote holistic and sustainable management of water resources (Moriarty et al. 2010).[1] However, taking into account that implementation of a comprehensive, integrated model may be impractical, recent applications have focused on applying IWRM principles at an appropriate level, from the international level to the community level (Butterworth et al. 2010). The methods selected should fit the given situation; inappropriate projects that do not result in useful outcomes can be

---

[A] Director, Center for Transboundary Water Management, Arava Institute; Co-authored by Tamee Albrecht, Natasha Westheimer, Jennifer Holzer.

[1] For more background on IWRM, see Global Water Partnership (GWP) 2000 and World Bank 2003.

counterproductive, leading to diminished confidence in governing institutions and a lost opportunity to tackle a problem (Butterworth et al. 2010; Corcoran et al. 2010).

Working within the political, social, and economic complexity of transboundary watersheds requires site-specific and adaptive wastewater management strategies. This is because up-stream and down-stream dynamics, differences in political and governance systems, and asymmetries in water and wastewater infrastructure make the implementation of centralized wastewater systems difficult. The hydrological reality is that watershed boundaries often do not follow political and administrative borders; in fact, 60% of the world's watersheds are transboundary, meaning that they cross international political borders. With 40% of the world's diverse population living in transboundary watersheds, water sources traverse areas of social and economic diversity—and occasionally politically conflicted areas, such as in Israel and the Palestinian Territories (UN Water 2008). A government mandated water management scheme is insufficient here because of this complexity, so we must reevaluate how to share water more effectively and promote cooperative water resources management. This is especially true where political conflict causes dramatic asymmetries in power and infrastructure, which is the case in the Israeli–Palestinian conflict as well as in other regions of conflict.

With little to no centralized sanitation services in the Palestinian West Bank and erratic centralized capacity in Gaza, communities and individuals rely on inadequate methods of waste disposal, often dumping untreated sewage directly into streambeds. Because the West Bank is most often up-stream from Israel, the untreated sewage blights Israel's attempts to maintain and protect groundwater resources,

streams, and public health. Further, this untreated sewage pollutes Israeli and Palestinian shared surface water and groundwater resources, which are already stressed by drought, population growth, and over-exploitation of a scarce supply. Pollution due to upstream sewage also exacerbates cross-border conflict thus, in these and other transboundary cases wastewater management and conflict resolution go hand in hand.

The asymmetry in the Israeli–Palestinian context is perhaps extreme but it represents the complexity of transboundary water and wastewater management. Whereas, in Israel, a relatively stable government and economy has provided extensive centralized infrastructure (over 90% of the population is connected to a centralized wastewater treatment facility (Al-Sa'ed and Al-Hindi 2010)) only 35% of the Palestinian population has access to adequate sanitation (Al-Sa'ed 2010). Further, the Palestinian Territories are dependent on financial support for wastewater infrastructure projects from international donors (Aburdeineh et al. 2010). In addition, political and bureaucratic barriers make it difficult for the Palestinian Authority to obtain permits from the Israeli government to build and operate centralized wastewater treatment facilities in the West Bank. For example, plans for a centralized wastewater treatment facility to serve Hebron were stalled for almost a decade and the facility has yet to commence service (Qudsi 2014). In such a situation, large quantities of untreated Palestinian sewage flow from the West Bank into Israel. It is estimated that as much as 60 mcm/year of untreated sewage crosses from the West Bank into Israel.

Drawing lessons from the Israeli–Palestinian context for transboundary wastewater management means one has to go beyond the accepted large-scale technical and often centralized approach to wastewater management. In transboundary watershed contexts, where a large proportion of the population is not being served by centralized sewage and wastewater treatment systems, a decentralized or off-grid approach may be more effective; on-site treatment for wastewater may provide both environmental and community benefits. Decentralized infrastructure solutions are therefore one important path toward improved wastewater treatment, especially where political complexity hampers effective cooperation and coordination between two or more national or cultural polities.

Under such scenarios, we advocate for a community-driven, decentralized, and flexible approach in those communities and regions where large-scale centralized solutions are inappropriate due to inadequate infrastructure, political difficulties, and funding constraints. Decentralized approaches to wastewater management include administrative decentralization and community infrastructure solutions. Administrative decentralization shifts decision-making and responsibility to lower level organizations, such as non-governmental organizations (NGOs) and user associations, thereby harnessing the capacity of community-based organizations and initiatives. Engaging stakeholders in decision-making, community education, and knowledge-sharing can produce more sustainable solutions that better reflect the needs and values of the users (Mody 2004). Community-based organizations like ours, the Arava Institute for Environmental Studies, a transboundary environmental NGO, often advocate the use of decentralized infrastructure alternatives. These are technologies that we, along with local partners and the users themselves, can install and monitor on a household or community scale. Decentralized solutions, such

as well-maintained septic systems or grey-water recycling systems, can be an effective and economical approach that avoids large capital investments and reduces operation and maintenance costs (USEPA 2005). The approach also precludes high-level political engagement as the systems are integrated at the local level. In many cases, this approach is more efficient than seeking high-level political sanction, especially when the polities lack a high-level cooperative mechanism for providing wastewater management services, as is the case in the Israeli–Palestinian context.

It is important, therefore, that these decentralized approaches be linked to conflict resolution. Consequently, stakeholder engagement is a key component of comprehensive transboundary watershed assessment, effective implementation of decentralized wastewater treatment solutions, and conflict resolution. During a transboundary watershed assessment, stakeholder participation that is inclusive of all parties must be integral to knowledge generation and data collection because in many cases an asymmetry exists in data collection, availability, and analysis. This type of stakeholder engagement must be fully representative of all parties within a transboundary watershed and thus will include stakeholders from different countries, regions, and communities. Stakeholders must also interact on a level playing field despite the power asymmetries that may exist among them. This is of course not simple to do but if the engagement is consistent and long-term, the relationships among the stakeholders should strengthen and issues of mistrust, and miscommunication should give way to cooperation. In regions of conflict, it is just these kinds of activities that may trickle up to the political sphere and offer solutions for high-level resolution of the conflict beyond the specific issues being addressed by the stakeholders themselves. The work of the Arava Institute attests to the axiom that in a region of conflict, relationships build trust and trust builds cooperation.[2]

Decentralized approaches are one viable path toward improved wastewater management in complex transboundary scenarios such as Israel and the Palestinian West Bank. In the Hebron/Besor watershed, one Israeli–Palestinian transboundary watershed, we are assessing pollution in the watershed and implementing decentralized greywater treatment infrastructure. Through this work, we have come to understand the importance of collaboration and face-to-face meetings for learning and improving wastewater management and project implementation in a region of conflict. Key lessons suggest the primacy of relationship building, stakeholder engagement, and maintaining an adaptive approach overall. We summarize our lessons learned here:

1. A decentralized approach to wastewater management is most appropriate in transboundary watersheds where there is political conflict and power asymmetries.
2. A community-based approach should be applied to transboundary watershed assessment, which can also promote information sharing and cooperation among stakeholders.
3. We recommend decentralized wastewater treatment infrastructure for low-income, rural areas and off-grid communities where on-site treatment and reuse of wastewater provides direct community benefit in a politically expe-

dient way especially when larger-level political complexity and conflict exists. Essential to the success of any such decentralized treatment project will be ongoing monitoring and capacity-building support to promote project sustainability.

Decentralized infrastructure, community-level management approaches, and horizontal information-sharing may achieve significant impacts in the short term, and begin to build a platform for expanded collaboration in the long term in those regions where water and other political conflicts are prevalent. We emphasize consulting local stakeholders and beneficiaries of projects to genuinely listen to and address their needs and concerns, before, during, and after project implementation. We call on other researchers to approach solutions to water issues by adapting methods to fit the sociopolitical context of the situation and drawing on the knowledge and capacities of the communities the solutions will serve. Doing so carries potential for increasing opportunities for collaboration, and finding solutions that actually last.

References

Aburdeineh, I., Bromberg, G., Michaels, L. and Theshner, N. 2010. "The Role of Civil Society in Addressing Transboundary Water Issues in the Israeli-Palestinian Context." In *Water Wisdom*, Ch. 13.

---

[2] For more information on the role of stakeholder engagement, please see J. Holzer, T. Albrecht, N. Westheimer, and C. Lipchin *anticipated 2014*. "Leveraging Environmental Data to Promote Cooperation toward Integrated Watershed Management in the Hebron/Besor Watershed." *Israel-Palestine Journal (accepted)*.

Al-Sa'ed, R. 2010. "A Policy Framework for Trans-Boundary Wastewater Issues along the Green Line, the Israeli-Palestinian Border." *International Journal of Environmental Studies* 67 (6): 937-954.

Al-Sa'ed, R. and Al-Hindi, A. 2010. "Challenges of Transboundary Wastewater Management along the Green Line-The Israeli-Palestinian Border." In *Shared Borders, Shared Waters: Israeli-Palestinian and Colorado River Basin Water Challenges*, eds. Sharon B. Megdal, Robert G. Varady, and Susanna Eden, First Edition. CRC Press-Balkema, Taylor & Francis Group, 203-220.

Butterworth, J., Warner, J., Moriarty, P., Smits, S. and Batchelor, C. 2010. "Finding Practical Approaches to Integrated Water Resources Management." *Water Alternatives* 3 (1): 68-81.

Corcoran, E., Nellemann, C., Baker, E., Bos, R., Osborn, D. and Savelli, H. eds. 2010. "Sick Water?" *The Central Role Of Wastewater Management In Sustainable Development*. A Rapid Response Assessment. United Nations Environment Programme, UN-HABITAT, GRID-Arendal.

Global Water Partnership (GWP) Technical Advisory Committee. 2000. Integrated Water Resources Management. Global Water Partnership, Stockholm, Sweden.

Holzer, J., Albrecht, T., Westheimer, N. and Lipchin, C. "Leveraging Environmental Data to Promote Cooperation toward Integrated Watershed Management in the Hebron/Besor Watershed." *Israel-Palestine Journal (accepted)*.

Massoud, M.A., Tarhini, A. and Nasr, J.A. 2009. "Decentralized Approaches to Wastewater Treatment and Management: Applica-

bility in Developing Countries." *Journal of Environmental Management* 90: 652-659.

Mody, J. 2004. "Achieving Accountability through Decentralization: Lessons for Integrated River Basin Management." World Bank Policy Research Working Paper 3346, June 2004.

Moriarty, P.B., Batchelor, C.H., Laban, P. and Fahmy, H. 2010. Developing a Practical Approach to 'light IWRM' in the Middle East." *Water Alternatives* 3 (1): 122-136.

Qudsi, S. (2014). *Introduction to House of Water and Environment Activities*. Ramallah: House of Water and Environment (HWE).

UN Water. 2008. Transboundary Waters: Sharing Benefits, Sharing Responsibilities. Thematic Paper. Task Force on Transboundary Waters.

U.S. Environmental Protection Agency (USEPA). 2005. Handbook for Managing Onsite and Clustered (Decentralized) Wastewater Treatment Systems: An Introduction to Management Tools and Information for Implementing EPA's Management Guidelines. EPA No. 832-B-05-001. December 2005.

World Health Organization (WHO). 2012. *UN-Water Global Annual Assessment of Sanitation and Drinking-Water (GLAAS) 2012 Report: The Challenge of Extending and Sustaining Services*. Geneva: WHO Press.

New Water Policy and Practice - Volume 1, Number 1 - Fall 2014

Documenting And Communicating Methods Used In Interdisciplinary Research And In Research Implementation

Gabriele Bammer[A]

In this opinion editorial, New Water Policy and Practice International Advisory Board member Prof Gabriele Bammer (National Centre for Epidemiology and Population Health, Australian National University, Australia) draws inspiration from her work in developing Integration and Implementation Sciences, a new discipline providing concepts and methods for conducting research on complex, real-world problems, and discusses the importance of documenting and communicating methods used in interdisciplinary research and in research implementation. Gabriele recognises that this is a key challenge of interdisciplinary research and proposes ways that contributors to New Water Policy and Practice Journal can help set the direction for interdisciplinary research.

Keywords: *interdisciplinary research; Integration and Implementation Science; policy and practice; contextual factors; research implementation*

A challenge for our new journal is to not only describe the results of our research on water policy and practice but also to share our methods and processes. Much of our research can be loosely described as "interdisciplinary". More specifically, it does one or more of the following:

- Brings together insights from different disciplines
- Includes perspectives from one or more stakeholder groups
- Seeks to influence change in water policy and/or practice.

One of the challenges of interdisciplinary research is that there are no agreed standard ways to describe the methods. Critical information about what was undertaken and how well it worked is inevitably missing from published accounts. This is not the case for established disciplines, such as chemistry or sociology, where methods are not only agreed and understood, but can be described in useful shorthand.

Interdisciplinary research tends to be more improvised, drawing on methods from a range of sources, adapting them as needed to the particular problem and creating new methods when established ones are not readily available. Describing such a process when there is no agreed framework can be both long-winded and 'clunky'. In such circumstances, the tendency is to highlight methods that are easily explainable and to gloss over or ignore more complex adaptation of methods and messy processes. But it is exactly here where insights often lie.

If we are going to build and refine our methods, we have to confront this dilemma. And we have to recognise that it will take time to develop agreed and polished ways of describing processes. Let us

---

[A] Australian National University, Australia

consider two suggestions for moving forward. One is the beginnings of a standard format for describing methods. The second is a way around the problem of long-windedness and clunky-ness.

Developing a standard format could start by asking authors to ensure they address the following five questions when they describe their methods. Each question would be asked both of the process of increasing understanding of the problem and of what happened when the researchers tried to influence policy and/or practice change.

1. What were the research and research implementation aiming to achieve and who were the intended beneficiaries?
2. What were the relevant 'components' (e.g. types of knowledge or groups targeted to influence change) and how were they chosen? This question includes:
   • The systems view taken of the research problem and of the policy/practice areas
   • The way they were scoped and boundaries set in deciding what to do and who to try to influence
   • The way issues were framed
   • How values, particularly conflicting values, were taken into account
   • How collaborations were managed.
3. How were the research and research implementation undertaken? For example, how were various kinds of knowledge brought together and how was influence exerted on policy/practice change? This involves describing methods for dialogue, modeling, communication, advocacy, engagement, and so on.
4. Were any contextual factors relevant? Here authors would reflect on the 'big picture' circumstances in which

the research and research implementation occurred, such as the political, economic, or historical context, any factors which worked for or against giving their research and research implementation legitimacy, and any institutional facilitators or barriers to the research or research implementation.
5. How well did the methods chosen work? This involves reflection on and evaluation of all the choices made in answering the previous four questions, for example, would a different systems approach have been more useful, what were the strengths and weaknesses of the dialogue method chosen, was the research helped by unrecognised contextual factors?

For more details, see Bammer (2013).

In time, as ways of thinking about and describing methods become more sophisticated, documenting processes at this level of detail will warrant journal articles of their own. In the meantime, *NWPP Journal* could copy the journal *Science* by making available an on-line 'additional materials' section, where messier processes and less straight-forward methods are described in detail. This section would be included in the peer-review. It could also be open for reader comment. Interaction between reviewers, readers and authors could lead to productive ways of enhancing methods and methods description.

Reference

Bammer, G. 2013. *Disciplining Interdisciplinarity: Integration and Implementation Sciences for Researching Complex Real-World Problems*. ANU E-Press. http://press.anu.edu.au?p=222171.